PLACES OF POWER

Hotspots in Colorado

A Guide to Harnessing Earth's Energy

Christine J. Dimon

Places of Power: Hotspots in Colorado

Copyright © 2025 c.d.
All rights reserved.

This is a work of nonfiction that explores both geological science and metaphysical interpretation. While the geological data has been researched and cross-referenced to the best of the author's ability, readers should consult additional sources for scientific applications. The energetic perspectives presented reflect personal exploration and are intended to inspire, not diagnose or prescribe.

First Edition
ISBN-13: 979-8-218-68286-6

Cover design and interior layout © 2025 by c.d.
Illustrations and energy signature icons © 2025 by c.d.
Published by Power Spots Press
Printed in the United States of America

For inquiries, collaborations, or rights requests, contact:
⌨ powerspotscolorado@gmail.com
◎ Instagram @powerspotscolorado,
or TikTok @powerspotscolorado

CONTENTS

SPIRITUAL PRACTICES AND ENERGY

PREFACE

Listening to the Land's Electric Whisper

Not all energy is visible. Some hums beneath the feet, in quartz veins that crackle with ancient stress. Some flows like water—through fault lines and clay seams, rising up in the quiet between heartbeats. Colorado, with its jagged peaks and deep valleys, doesn't just offer beauty—it pulses with a mysterious charge.

This book is born from that pulse.

Places of Power invites you to explore Colorado through a dual lens: geophysical and metaphysical. On one level, we'll examine the hard science—conductivity discontinuities, telluric currents, piezoelectric effects, and magnetic anomalies. On another, we'll feel into what these places do to the human body and mind. How some spots make the air feel electric. How meditating on a quartzite outcrop can stir visions or clarity. How sunrise over a geothermal spring, shifts not just temperature—but perception.

This is not a guidebook in the traditional sense. It's a field manual for those seeking a deeper experience of place. Each location we visit is presented with geological context, energetic signature, and practical tips for tuning in— whether through meditation, yoga, breathwork, or simply listening.

You'll also find reference notes in the margins—fast facts, key terms, and theories that connect the subtle and the substantial. Whenever possible, we explore both scientific and symbolic interpretations: why a fault zone might both channel electric currents and spiritual awakening. Why an ore-rich zone might amplify geomagnetic activity and human intuition.

Whether you're a geologist curious about how conductivity and magnetism shape landscapes—or a seeker chasing the currents beneath your own skin—this book is for you.

Let it be your companion into Colorado's energetic underworld. Let it sharpen your perception, deepen your practice, and electrify your connection to the Earth.

– c.d.

ACKNOWLEDGMENTS

This book was born from a lifelong fascination with the way energy moves—through earth, through rock, through people.

To the teachers who first introduced me to geology and the ones who later taught me how to feel the earth instead of just measure it—thank you. Your guidance made it possible to bridge science and soul on every page.

To those who walked these places of power with me, in silence and in laughter—your presence grounded this work. You reminded me that energy is not just a concept; it's a conversation.

To Colorado, whose ancient folds, magnetic quirks, and fault-line dreams shaped every chapter—I am endlessly in awe.

To the friends, readers, and kindred spirits who believed in this book before it had a name—you helped it come alive.

And finally, to the currents themselves—the telluric flows, the piezoelectric sparks, the morning stillness charged with something unnamed—thank you for showing me that sometimes, the Earth speaks first.

HOW TO USE THIS BOOK

This book is not just meant to be read—it's meant to be felt, walked, and lived. Places of Power is your guide to Colorado's most energetically charged landscapes—locations shaped by geologic intensity, electromagnetic complexity, and spiritual significance. These are places where the Earth speaks more clearly, and where the body and spirit are more likely to listen.

Each chapter explores one of these powerful sites in depth, offering a balance of geological insight, metaphysical interpretation, and experiential guidance. You'll find detailed write-ups, energetic profiles, historical notes, and reflections on how these places can shift your awareness, well-being, and connection to the Earth.

To get the most out of your journey—whether you're reading from home, hiking the trail, or meditating by a fault line—here are some tools and tips to orient you:

Understanding the Margins

Throughout the book, you'll notice margin guides and callouts—these are your field notes and fast insights, designed to be quick to reference and powerful in practice:

◆ ENERGY SIGNATURE GUIDE: A standardized profile that gives you quick info about a location's energetic essence, including:

📍 Location
⌛ Peak Energy Time
⚡ Energy Type
🧘 Best Activity
🎯 Symbol for the site's energetic archetype

⚒ Geological Tips: Concise notes explaining the site's rock types, fault activity, magnetic or conductive properties, and why it matters energetically.

🌐 Energy Insights: Spiritual and scientific sidebars that explore electromagnetic phenomena, bioelectric effects, and how the human body may interact with these fields.

🧭 Ley Line Notes: Where relevant, we note alignments with ancient pathways, energetic intersections, or symbolic crosspoints.

These margin notes are ideal for quick reads before arriving at a site or for grounding your experience while you're there.

🕐 When to Visit for Maximum Energy

Each location in this book has its own energetic rhythm, and natural cycles matter. Where possible, we've included the Peak Energy Times based on:

- and lunar alignments (solstices, Solar equinoxes, new and full moons)
- Time of day (sunrise, sunset, high noon, or midnight)
- Seasonal shifts (spring melt, autumn stillness, post-storm clarity)
- Weather conditions (air ionization, storm energy, mist, or wind)

Use these cues to time your visit for resonance—not just scenery. When your breath aligns with the landscape's charge, transformation becomes accessible.

🧘 How to Work with the Energy

Every chapter includes recommended energetic practices to help you tune in. These may include:

- Grounding rituals (barefoot walking, breath-holds, contact with rock or soil)
- Meditation styles (stillness, movement, sound, visualization)
- Breathwork suggestions (timed breathing, wind alignment, floral breath)
- Chakra alignments based on the landscape's frequency

- Symbolic or archetypal associations to help guide intention setting

This book encourages you to feel your way through, to work with intuition as much as intellect.

Scientific and Spiritual Integration

You do not need to choose between geology and mysticism. This book is built on the idea that the two are in conversation.

- Quartz can both generate electrical charge and amplify clarity of thought.
- Fault zones can conduct current and open symbolic passageways.
- Water can erode rock and cleanse the emotional field.

Feel free to use whichever lens works for you—or better yet, try both.

Planning Your Journey

While this book can be used for armchair exploration, it is best experienced outdoors. You might consider:

- Building your own power place pilgrimage, visiting a handful of sites in a seasonal loop
- Using the symbols to track how each place affects you emotionally or energetically
- Revisiting sites at different times of year to see how energy shifts
- Keeping a field journal or energetic log as you move through the book

Final Note

This is a book meant to be opened again and again. Whether you're planning a weekend getaway, seeking healing, or simply curious about how Earth's dynamic power shapes your inner world, this guide is here to walk with you.

May it remind you that the Earth is not just beneath you—it is within you, speaking in stone, water, charge, and silence.

NATURE-CONNECTED & ENERGETIC

Harnessing the energy of water, earth, and breath.

INTRODUCTION

DISCOVER HOW EARTH'S ENERGY HOTSPOTS CAN TRANSFORM YOUR LIFE

Hidden among Colorado's jagged peaks, vast valleys, and high desert landscapes lies a secret force humming beneath the surface. These "places of power" aren't just pretty vistas—they're hotspots of pure, untamed energy. These zones blur the line between science and the spiritual, opening portals to a deeper connection with the Earth's heartbeat—and your own.

For centuries, seekers of knowledge, energy, and enlightenment have been drawn to these places, sensing something beyond the visible—a resonance, a pulse, a force that cannot be ignored. The ancient Indigenous tribes of Colorado recognized these lands as sacred, attributing them with spiritual power and divine connection. Modern science has started catching up, uncovering the electromagnetic anomalies, conductivity discontinuities, and telluric currents that shape these sites. But the true essence of these places lies in the experience—the moment you step onto charged ground and feel something shift within you.

What makes these locations so special? The answer lies deep in the Earth itself. Colorado's landscapes are a playground of geological wonders, from towering quartz-rich peaks that generate piezoelectric effects under stress to volcanic fields brimming with geothermal energy. These sites are more than just beautiful; they are alive with energy, pulsing with a natural force that influences not only the environment but also human perception, physiology, and spiritual awareness.

Imagine standing at the edge of the San Luis Valley, where electromagnetic anomalies dance through the air, stirring a deep awareness within you. Or wading into the steaming waters of Mount Princeton Hot Springs, where electrically charged minerals soak into your skin, leaving you revitalized. Or walking through the towering red rock formations of Garden of the Gods, where the land seems to hum, whispering ancient secrets to those who take the time to listen.

This book is your guide to uncovering these enigmatic locations, understanding the science and spirituality behind them, and harnessing their potential for personal transformation. Whether you seek adventure, healing, or a deeper connection to the unseen forces of the Earth, Colorado's power spots have something to offer.

Are you ready to tune in, step forward, and let the energy of the land guide you? Let's begin the journey.

THE SCIENCE BEHIND "PLACES OF POWER"

ROCKS THAT BUZZ WITH LIFE AND LORE

Understanding how Earth's energy hotspots can transform your life

The Earth isn't just dirt and stone—it's alive with energy. Colorado's unique geology, from quartz-studded peaks to ancient volcanic fields, creates the perfect recipe for a natural symphony of electromagnetic and geothermal forces that captivate scientists and spiritual seekers alike.

Beneath Colorado's towering peaks and rolling valleys, a vast network of mineral-rich formations hum with geologic vitality. Quartz, a dominant mineral in many of the state's formations, is known for its piezoelectric properties—meaning it generates an electrical charge when subjected to mechanical stress. This natural electricity pulses beneath the surface, aligning with deep geological structures and fault lines, forming energetic hotbeds that have long been recognized by Indigenous cultures as sacred places. These energy fields are not just folklore—they are backed by scientific observations showing that shifts in the Earth's crust can generate detectable electromagnetic anomalies.

Did you know that
Pikes Peak granite has piezoelectric properties?

Colorado's volcanic history also plays a crucial role in shaping its unique energetic signature. Ancient eruptions left behind vast deposits of igneous rock, rich in metallic minerals that enhance conductivity and influence local magnetic fields. When combined with the state's numerous geothermal springs, these deposits create environments where heat, pressure, and mineral interactions form a constant flow of subtle electrical currents. These dynamic interactions create landscapes where energy flows through the ground, manifesting as heightened electromagnetic activity.

One of the most fascinating aspects of Colorado's geology is how different rock types interact to influence energetic sensitivity. High-quartz-content formations like the Rocky Mountains amplify piezoelectric effects, whereas sedimentary layers rich in organic materials, such as the Denver and Piceance basins, contribute to natural conductivity. This unique geological layering makes some areas more receptive to telluric currents—natural electric currents flowing through the Earth's crust. These currents, influenced by changes in solar and lunar activity, create power spots where the land itself seems to radiate with an unseen force.

Fault zones further enhance Colorado's energetic landscape. These fractures in the Earth's crust act as natural conduits for energy, focusing electromagnetic fields and telluric currents into specific locations. When these fault zones intersect with quartz-rich formations or geothermal activity, they create highly charged power spots—places where people often report sensations of warmth, tingling, or deep mental clarity. Scientific instruments can measure these effects, with studies revealing measurable variations in the Earth's magnetic field and conductivity in these areas.

Historically, these power spots have been revered by cultures seeking harmony with the Earth's natural energy. From the Ancestral Puebloans to modern spiritual seekers, people have long been drawn to locations where the Earth's electromagnetic activity aligns with human consciousness. Whether through meditation, rituals, or simply being present in these environments, individuals report profound experiences of grounding, expansion, and heightened awareness.

The connection between geology and consciousness is not just theoretical. Research into the brain's response to electromagnetic fields suggests that exposure to certain frequencies can alter mental states, promoting relaxation, creativity, and heightened intuition. Some of Colorado's power spots may act as natural amplifiers of these effects, explaining why so many people feel deeply affected by the land's energy.

By understanding the geological forces at play, we can begin to map these power spots with greater precision. Utilizing tools like magnetometers and resistivity surveys, scientists and enthusiasts alike can uncover areas where energy accumulates and shifts, creating tangible locations for those seeking to connect with the Earth's vibrational essence. The interplay between rock type, tectonic movement, and conductivity forms an intricate web that underlies the mysterious, compelling energy that Colorado's landscapes exude.

This alchemy of geology and energy is what makes Colorado a beacon for those attuned to the power of the land. Whether one experiences it through the deep silence of the mountains, the charged atmosphere of geothermal springs, or the inexplicable clarity of thought in certain landscapes, the energy here is undeniable. The question remains: are you ready to listen to what the Earth is telling you?

AND MAGNETIC FIELDS THAT PULSE WITH MYSTERY

*How the Earth's shifting magnetic field
influences emotions, meditation, and manifestation*

Where rock types collide, conductivity anomalies emerge like cosmic energy gates. These electric hotspots pulse with telluric currents and magnetic fluctuations, creating the perfect storm for anyone tuned into their surroundings. Discover the science behind this electrifying phenomenon and its mesmerizing effects on human perception.

Conductivity discontinuities occur where geological formations of varying electrical resistivity intersect. These regions act as natural energy focal points, directing electrical currents and shaping the magnetic environment in profound ways. When highly conductive minerals such as copper, iron, or graphite are present, they create natural circuits, influencing the surrounding electromagnetic field. Fault lines, mineralized zones, and geothermal hotspots serve as amplifiers of this energy, making them particularly attractive to those sensitive to Earth's energetic pulse.

Conductivity Discontinuity
*A sudden change in the Earth's ability to conduct electric currents,
often found at fault lines.*

Telluric currents, also known as Earth currents, flow through the planet's crust, influenced by solar activity, lunar cycles, and geomagnetic fluctuations. These electrical currents travel through conductive rock layers and groundwater channels, further energizing areas where conductivity discontinuities are most pronounced. The movement of these currents can change with the time of day, season, or atmospheric conditions, making some power spots more active at specific moments. The ancient builders of sacred sites may have intuitively recognized these fluctuations, aligning their temples and stone circles to harness and amplify these energies.

Magnetic field variations are another key component of electrically active regions. Geological formations rich in ferromagnetic minerals, such as magnetite, create local magnetic anomalies that can subtly influence human perception and cognition. Studies have shown that exposure to strong magnetic fields can alter brain wave activity, inducing heightened awareness, relaxation, and even altered states of consciousness. This might explain why certain locations, such as the Garden of the Gods or the San Luis Valley, are often reported as places of deep spiritual resonance and unusual experiences.

Some of the most fascinating effects of conductivity discontinuities stem from their ability to create localized electric fields. When mechanical stress—such as tectonic activity or even foot traffic—interacts with quartz-bearing formations, a piezoelectric effect occurs, generating electrical charges that can further amplify the energy of a site. These piezoelectric discharges are believed to contribute to the luminous phenomena often reported in highly active areas, such as orbs of light, unexplained flashes, or even sensations of tingling energy.

Colorado's diverse geology makes it an exceptional location for experiencing these electromagnetic interactions. From the uplifted fault zones of the Front Range to the deeply buried mineral veins of the San Juan Mountains, the state is crisscrossed with energetic pathways that shape both the land and the human experience. **Hot springs, for instance, are often found in areas with significant conductivity discontinuities, as the movement of mineral-rich water creates an ideal conduit for telluric currents.** Similarly, high-altitude locations with strong solar exposure can enhance ionization in the air, heightening the electrical charge of a landscape.

The relationship between conductivity discontinuities and magnetic anomalies offers a compelling explanation for the mystical qualities associated with certain locations. While modern instruments can measure these fields, ancient cultures relied on intuition and observation to identify areas of heightened energy. Today, those seeking to tap into these power spots can use scientific tools alongside personal experience to uncover the unseen forces at play.

As research into Earth's natural electromagnetic phenomena continues, we gain a deeper appreciation of how these geological interactions shape our world. Whether through meditation, scientific study, or pure curiosity, understanding the mechanisms behind conductivity discontinuities and magnetic fields brings us closer to the energy of the Earth itself. Those who take the time to explore these electric landscapes may find themselves forever changed by the pulse of the planet beneath their feet.

THAT THRILL THE LANDSCAPE AND ENERGIZE THE SPIRIT

The science of geoelectric activity and its effect on the body and mind

---

Water moves, rocks grind, and the Earth's magnetic field dances in perfect harmony. These forces unleash electric currents, transforming ordinary ground into a power grid for both body and soul. Dive into the intricate interplay that powers these enigmatic zones.

Electric currents within the Earth's crust are driven by a complex interaction of geological, atmospheric, and solar forces. These currents, known as telluric currents, are naturally occurring electrical flows that pulse through the planet's conductive materials. Areas where these currents concentrate tend to coincide with some of the most energetically active places on Earth, creating hotspots of unusual magnetic and electrical phenomena.

Water plays a key role in this process, acting as both a conduit and amplifier of natural energy. Flowing groundwater, particularly in mineral-rich formations, enhances conductivity and generates electromagnetic activity. Colorado, with its abundant geothermal springs, aquifers, and underground streams, is a prime location for this phenomenon. The mineral content of these waters—including silica, lithium, and iron—further influences conductivity, shaping the electromagnetic landscape of the region.

Tip
If you feel "off" or overstimulated in high-energy areas,
ground yourself by placing your bare feet on soil or rock.

Tectonic activity also contributes to the movement of electrical currents. When the Earth's plates shift, mechanical stress is applied to quartz-bearing rocks, triggering the piezoelectric effect—a phenomenon where electrical charge accumulates in response to pressure. This creates localized electric fields that can influence both the environment and human perception. Many of Colorado's most well-known power spots, such as the Rocky Mountain foothills and geothermal basins, are located along fault lines where tectonic forces generate high energy activity.

Geomagnetic fluctuations add yet another layer of complexity. The Earth's magnetic field is in constant motion, shaped by solar winds and the planet's molten core. When this field interacts with the conductive materials in the ground, it induces additional electrical currents. These interactions can intensify at specific times of day or during solar storms, temporarily increasing the energetic charge of a landscape. Locations with large deposits of magnetite, hematite, or other iron-rich minerals tend to experience the strongest magnetic anomalies, amplifying the sense of energetic intensity.

Certain frequencies are associated with relaxation, heightened awareness, and even altered states of mind. This could explain why so many visitors report unusual sensations, vivid dreams, or deep meditative states when spending time at power spots in Colorado.

Some ancient cultures appeared to have an intuitive understanding of these energy flows. Indigenous tribes in Colorado often built sacred sites in locations where electric currents were naturally concentrated. The alignment of stone circles, petroglyph sites, and ceremonial spaces suggests an awareness of the land's energetic properties long before modern scientific measurements could confirm their existence.

Today, modern explorers can use tools like electromagnetic field meters and magnetometers to detect these invisible forces. Yet, the most powerful tool remains personal experience. Whether standing near a thundering waterfall, hiking along a quartz-streaked ridge, or resting beside a bubbling geothermal spring, the presence of the Earth's electric currents is something that can be felt as much as measured.

By understanding the science behind these natural phenomena, we can appreciate how energy moves through the land, shaping both the physical and spiritual landscapes of Colorado. Those who seek out these electrified places often describe a sense of renewal, clarity, and connection to something greater than themselves. It is here, in the charged embrace of the Earth's dynamic currents, that we find the pulse of the planet itself, waiting for those who are willing to tune in.

AND THE ENERGY OF ANCIENT STRESS

How the immense forces of mountain-building events
leave behind geological scars that still resonate with energy

The concept of foreland basins and their sedimentation patterns can enhance
the scientific depth of our understanding of Colorado's power spots by linking
their geological history to the way energy moves through the Earth's crust.

Geological Memory and Energy Flow - Foreland basins record long periods of stress and compression as mountains form. These stress zones can store and later release energy—both mechanically (earthquakes, fault reactivations) and electromagnetically (piezoelectric effects in quartz-rich rocks). Colorado's Rocky Mountain foreland basin developed due to Laramide compression. Could areas with high conductivity, such as faulted zones along ancient foreland structures, be energetically active? Some power spots may be located where ancient stresses are still subtly influencing the crust, creating conductivity discontinuities.

Geological Tip
Formed under tectonic pressure, foreland basins
may amplify piezoelectric effects in quartz-rich layers.

Foreland Basin Sediments and Energy Sensitivity - These basins often contain a mix of clastic sediments, particularly sandstones and shales. Sandstones (high quartz content) can enhance piezoelectric effects, while shales (rich in organic matter) might influence localized conductivity. If a power spot sits on or near these materials, there may be a link between rock type and energy sensitivity—similar to how quartz is said to amplify energy in metaphysical traditions.

Ancient Sedimentary Deposits and Modern Energy Features - Some of the best power spots might sit on buried ancient foreland basin deposits where groundwater movement, mineral composition, or old stress regimes enhance the geophysical environment. Since these sediments were deposited ahead of thrust sheets, the tectonic forces at play could have restructured the subsurface, creating pathways for electrical currents.

Foreland Basins and Magnetic Anomalies - Some regions of foreland basins have gravitational and magnetic anomalies, possibly due to deep crustal features, mineral concentrations, or groundwater interactions. Could certain meditation sites in Colorado be located where these foreland structures still influence the geomagnetic field?

By mapping foreland-related features (ancient thrust belts, sedimentary thickness variations) onto power spot locations, we can see if there's a correlation between the geological past and modern energy experiences. Using real-world geophysical data, like magnetotelluric surveys or gravitational anomaly maps, we can scientifically support the idea that Colorado's energy landscape is shaped by its foreland basin history.

INTERACTIVE MAP ACCESS

Explore Colorado's Energetic Landscapes in Real Time

Welcome to the Living Map of Places of Power

This book is just the beginning.

To help you deepen your connection with Colorado's energetic hotspots, I've created a custom interactive map where you can:

📍 Locate sacred and geoelectric sites

🧭 Plan high-energy visits based on peak times

📖 Read Energy Signature Guides on the go

💫 Track your own journeys across seasonal and solar cycles

Map Features Include

📍 **Hotspot Markers** — Each location includes its Energy Signature, best times to visit, geological notes, and meditation suggestions.

🏔 **Seasonal Timing Tips** — Learn when sunrise, solstice, or storm energy will amplify a site's field.

⚡ **Geology + Energy Integration** — Understand how rock type, fault zones, and water flow enhance the energy you feel.

🧘 **Recommended Field Practices** — Suggested meditations, breathwork, rituals, and intuitive exercises tailored to each spot.

🏔 **Seasonal Tips** for different times of year.

 ## Use This Map To

✨ Plan sacred solo hikes

🌿 Organize seasonal pilgrimage loops

🔋 Discover spontaneous side-trails to additional energetic sites

🧍 Stay connected to the pulse of the land wherever you roam

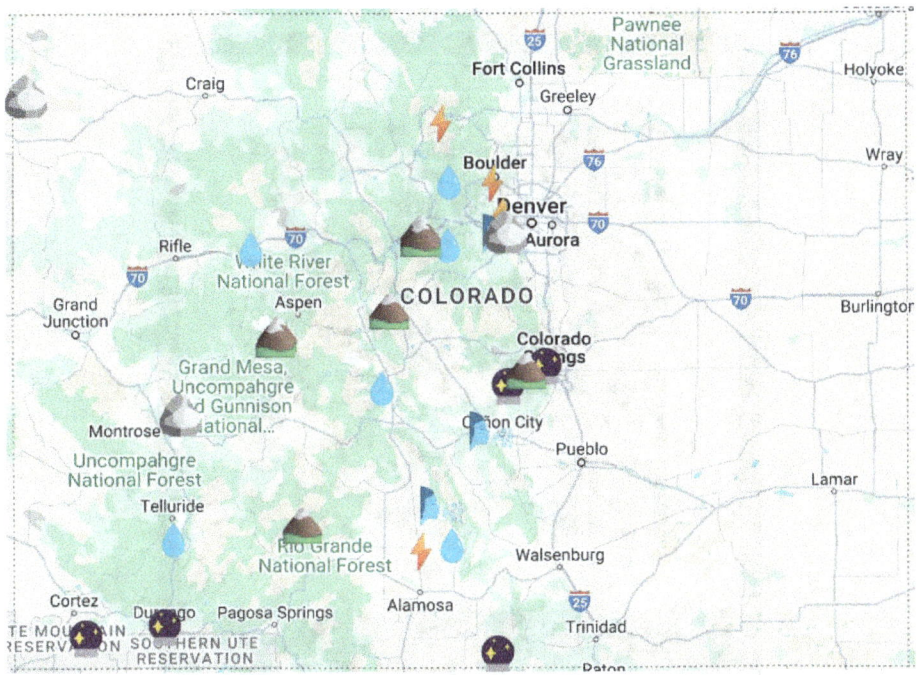

A Living Map: Updated with the Earth's Pulse

This map will grow.

New discoveries, reader experiences, and seasonal notes will be added over time—because **Earth's energy is not static. It's alive, just like you.**

Bookmark it. Wander it. Let it be your **compass back to resonance**.

Scan. Explore. Remember.
Scan this QR Code to open the map instantly

SCAN ME

or
Visit: https://tinyurl.com/mwxzt3jj

"The Earth does not speak in words—it speaks in paths."

ELECTRICALLY ACTIVE LOCATIONS IN COLORADO

MOUNT PRINCETON HOT SPRINGS

LIQUID FIRE BENEATH THE EARTH, A GEOTHERMAL HAVEN

Imagine soaking in mineral-rich, electric-charged waters while the energy of the Earth seeps into your being. That's Mount Princeton—a geothermal marvel igniting your spirit and rejuvenating your mind. Explore its rich history and why it's a magnet for energy seekers.

◆ ENERGY SIGNATURE GUIDE ◆

Location: Mt. Princeton Hot Springs

Peak Energy Time: Sunrise, Spring & Autumn Equinoxes

Energy Type: Thermal currents, ionized air, & grounding potential from geothermal water

Best Activity: Deep meditation while submerged in hot springs

 Symbol: Triple Wave Lines (Geothermal flow & movement)

Nestled in the Collegiate Peaks of central Colorado, Mount Princeton Hot Springs is more than just a place to relax—it is a site of immense geological power. Fed by natural geothermal springs that surge from deep underground, this location offers a unique blend of healing, history, and earth energy that has drawn people for centuries. The springs emerge at temperatures exceeding 120°F, their mineral-rich waters carrying the vitality of the Earth's molten core straight to the surface.

The geothermal activity at Mount Princeton is fueled by a deep fault system that allows heat from the Earth's interior to rise through cracks in the crust. As groundwater seeps downward, it is heated by the geothermal gradient before rushing back up to the surface, infused with dissolved minerals such as lithium, calcium, and magnesium. These elements not only enhance conductivity but are also believed to promote healing, relaxation, and spiritual clarity.

Long before modern visitors came seeking wellness retreats, Indigenous tribes, including the Ute people, revered these hot springs as sacred sites. They believed that the waters carried the Earth's wisdom and could restore balance to both body and mind. Even today, many who visit report experiencing a deep sense of renewal, as if the springs wash away not only physical tension but also energetic blockages.

Scientific studies have shown that geothermal waters can generate subtle electrical fields due to their high ion content. When bathers immerse themselves in these waters, they are not just absorbing heat but also interacting with natural electromagnetic energy. Some researchers suggest that this could influence the nervous system, potentially enhancing meditation, creativity, and emotional well-being. This phenomenon aligns with ancient traditions that regard hot springs as portals to heightened states of awareness.

Beyond its energetic properties, Mount Princeton Hot Springs sits in a landscape teeming with geological features that contribute to its unique vibrational qualities. The surrounding mountains are composed of granite, a rock type known for its high quartz content. Quartz, with its piezoelectric properties, can generate an electrical charge when subjected to pressure, further amplifying the energetic potential of the area. Visitors often describe a tingling sensation while soaking in the springs—a possible result of the interaction between their body's own bioelectric field and the geothermal waters.

Whether you visit to soothe aching muscles, meditate in the warm mineral pools, or simply immerse yourself in the power of the Earth, Mount Princeton Hot Springs is a place where the line between science and spirituality blurs. It is a reminder that beneath the surface, the planet is alive, constantly shifting, pulsing, and radiating its ancient energy for those who are open to receiving it.

STONE TITANS AND THEIR FREQUENCIES
THAT HUM WITH POWER

These ancient red sandstone formations don't just stand tall; they hum with vibrations that inspire awe. Their alignment with magnetic anomalies makes them an energy seeker's dream, a living monument to Earth's power.

◆ ENERGY SIGNATURE GUIDE ◆

Location: Garden of the Gods, Colorado Springs

Peak Energy Time: Sunrise, Equinoxes, and After Thunderstorms

Energy Type: Quartz-rich sandstone, residual tectonic stress, atmospheric ionization

Best Activity: Walking meditation & grounding breathwork among the fins

Symbol: Red Rock Spire – Represents upright energy flow and ancestral connection

Garden of the Gods, located in Colorado Springs, is one of the most striking and energetically potent landscapes in the state. Towering slabs of red and white sandstone rise from the ground, forming jagged spires that seem to defy gravity. These formations, sculpted over millions of years by geological upheaval and erosion, create an environment charged with electromagnetic energy that has long been recognized by indigenous cultures and spiritual seekers.

At the heart of this energetic marvel is the unique mineral composition of the rock formations. The red sandstone, composed primarily of quartz and feldspar, exhibits piezoelectric properties, meaning it can generate electrical charge when subjected to mechanical stress. This phenomenon may contribute to the tangible sense of energy that many visitors report when walking through the park. Combined with the site's proximity to fault lines, which act as conduits for natural electrical currents, the area is an energetic vortex that amplifies subtle earth energies.

Geologically, the formations are remnants of ancient sedimentary layers, deposited during the Paleozoic era when this region was covered by vast inland seas. Over time, tectonic forces uplifted these layers, tilting them into their present vertical positions. This upheaval not only created the dramatic visual landscape but also introduced fractures and fault lines that enhance conductivity, making Garden of the Gods an electromagnetic hotspot.

Magnetic anomalies have been detected in the area, possibly caused by the concentration of iron-rich minerals within the sandstone. These fluctuations in the Earth's magnetic field can subtly influence human physiology, altering brain wave activity and promoting states of relaxation, heightened awareness, and deep contemplation. Some visitors report feelings of euphoria or an inexplicable sense of connection to the land—experiences often attributed to the site's energetic properties.

Indigenous tribes, including the Ute and Cheyenne, recognized the spiritual power of this place long before modern scientific investigations. They held ceremonies among the towering stones, believing the formations to be inhabited by ancestral spirits and protective forces. Legends tell of warriors receiving visions and guidance from the land, reinforcing the idea that this sacred space is a conduit for higher consciousness.

🧭 Ley Line Note

Garden of the Gods is believed to sit at the junction of ancient ley lines— connecting Pikes Peak, the Sangre de Cristo Mountains, and Chimney Rock. Some say this crossroads amplifies intuition and ancestral memory.

Beyond its energetic and spiritual significance, Garden of the Gods serves as a natural sanctuary for grounding and rejuvenation. The vast open sky, striking rock formations, and the interplay of light and shadow create an environment that encourages mindfulness and presence. Many visitors practice meditation or yoga among the formations, seeking to harmonize their own energy with the Earth's pulse.

Whether you are drawn to the park for its breathtaking beauty, its scientifically fascinating electromagnetic properties, or its deep spiritual resonance, Garden of the Gods stands as a testament to the power of the natural world. It is a place where the veil between the physical and energetic realms feels thinner, inviting all who visit to tune into the subtle frequencies of the Earth and rediscover their own connection to something greater.

LOOKOUT MOUNTAIN

WHERE NATURE'S PULSE MEETS THE SKY AND EXPANDS AWARENESS

Lookout Mountain isn't just a vantage point; it's a confluence of geological energy fields and atmospheric brilliance. Here, Earth and sky unite to spark inspiration and an elevated sense of being.

◆ ENERGY SIGNATURE GUIDE ◆

📍 **Location:** Lookout Mountain, Golden, CO

⏳ **Peak Energy Time:** Sunset, Autumn Equinox, and during temperature inversions

⚡ **Energy Type:** Fault-aligned quartz veins, telluric currents from water-saturated clays, and atmospheric charge differentials

⚗ **Best Activity:** Sunset meditation, energy clearing rituals, and vision journaling

🎯 **Symbol:** 🏔 Mountain Overlook – Represents expanded perception and horizon awareness

Perched high above Golden, Colorado, Lookout Mountain offers breathtaking views of the Front Range and Denver below. But beyond its scenic appeal, this site pulses with subtle yet undeniable energy, making it a destination for those drawn to places of heightened awareness. The mountain's position along a geological fault line and its rich mineral composition contribute to its natural conductivity, amplifying telluric currents that flow through the region.

Lookout Mountain sits at the intersection of powerful earth forces. The mountain's foundation is composed of ancient metamorphic rock, including quartz-rich formations that exhibit piezoelectric properties. When mechanical stress is applied—whether from tectonic activity or even shifts in atmospheric pressure—these quartz formations generate electrical charges. This effect may explain the sensations of clarity and upliftment reported by visitors who come to meditate, reflect, or simply soak in the mountain's unique energy.

⚒ *Geological Tip*

The Golden Fault runs beneath Lookout Mountain, channeling groundwater and stress through quartz veins. This may create subtle electromagnetic activity—ideal for grounding, clarity, and emotional reset.

Historically, indigenous peoples revered Lookout Mountain as a sacred space, recognizing its significance as a spiritual threshold between the terrestrial and celestial realms. It was a place for vision quests, ceremonies, and connecting with higher consciousness. The mountain's commanding presence and its alignment with solar and lunar cycles further reinforce its role as an energetic conduit.

Modern science supports the notion that high-altitude locations with significant electromagnetic activity can influence human perception. Variations in the Earth's magnetic field at Lookout Mountain, possibly due to the concentration of iron-rich minerals, can subtly affect brainwave activity. This could enhance states of meditation, creativity, and intuitive insight, making it a preferred location for those seeking clarity and spiritual expansion.

The interplay between geological formations and atmospheric conditions also adds to Lookout Mountain's energetic appeal. The high elevation creates an environment where the air is thinner and more ionized, leading to increased oxygen intake and heightened sensory perception. Visitors often describe feeling lighter, more present, and deeply attuned to their surroundings. The way sunlight dances across the rock formations and the crisp mountain air contribute to the site's revitalizing effect.

For many, Lookout Mountain serves as a place of transition—a space to release burdens, set new intentions, or simply reconnect with the rhythms of the natural world. Whether watching the sunrise cast golden hues over the landscape or standing beneath a sky filled with stars, the mountain has a way of shifting awareness and expanding perspectives.

The convergence of geology, atmosphere, and subtle energy fields makes Lookout Mountain a powerful destination for seekers of insight and renewal. It is a reminder that places of power are not merely defined by their physical beauty, but by the resonance they create within those who visit. To stand atop Lookout Mountain is to witness the fusion of Earth and sky, a meeting point of forces that invite reflection, inspiration, and an elevated state of being.

THE SAN LUIS VALLEY

GATEWAY TO THE MYSTICAL UNKNOWN
AND COSMIC PHENOMENA

A landscape of legends and mysteries, the San Luis Valley's electromagnetic phenomena attract more than stargazers. It's a portal to the unexplainable, where science and spirituality intersect in breathtaking ways.

◆ ENERGY SIGNATURE GUIDE ◆

Location: San Luis Valley, Southern Colorado

Peak Energy Time: Midnight, Equinoxes, and during geomagnetic storms

Energy Type: Telluric currents from the Rio Grande Rift, magnetic anomalies, geothermal flow, and atmospheric ionization

Best Activity: Stargazing meditation, dreamwork, and deep stillness practices

Symbol: Celestial Portal – Represents the thin veil between Earth and cosmos

---

The San Luis Valley is a vast, high-altitude basin stretching across southern Colorado and northern New Mexico. Encircled by the rugged peaks of the Sangre de Cristo and San Juan Mountains, this valley is one of the most enigmatic regions in North America. Known for its unusual energy signatures, unexplained lights, and a long history of UFO sightings, the valley has captivated scientists, spiritual seekers, and curious travelers alike.

One of the valley's most striking features is its strong electromagnetic activity. Local residents and researchers have documented fluctuations in the magnetic field, strange interference with electronic devices, and even reports of unexplained physiological effects. Some attribute this to the region's underlying geology—thick layers of sedimentary rock rest atop deeply buried fault lines, where geothermal and tectonic forces may generate telluric currents. These natural electric fields could amplify existing energy anomalies, creating an environment that heightens perception and awareness.

The valley's history of unexplained aerial phenomena is another layer of its mystique. Reports of glowing orbs, fast-moving objects, and unidentifiable crafts date back centuries, with indigenous Ute and Navajo tribes telling stories of sky beings and portals to other realms. The modern UFO movement has embraced the San Luis Valley as one of the most active hotspots for sightings in the United States. The presence of the UFO Watchtower, a site dedicated to monitoring the skies, reflects the region's deep connection to cosmic mysteries.

Beyond extraterrestrial encounters, the San Luis Valley is also home to numerous spiritual and metaphysical sites. Hot springs bubble up from the Earth, their mineral-rich waters believed to hold healing properties. Great Sand Dunes National Park, with its ever-shifting landscapes, exudes an almost otherworldly energy. Meanwhile, Crestone, a small town nestled at the base of the Sangre de Cristo Mountains, has become a haven for spiritual communities, with Buddhist centers, Hindu ashrams, and energy vortexes drawing seekers from around the world.

The valley's unique energy profile makes it an ideal location for meditation, vision quests, and deep introspection. Many visitors report experiencing an altered sense of time, heightened intuition, and an overwhelming sense of peace when spending extended periods in the region. Whether this is due to the valley's powerful electromagnetic fields or its rich spiritual history, the effects are undeniable.

Scientists continue to investigate the anomalies of the San Luis Valley, yet definitive explanations remain elusive. Is it an interdimensional gateway, a natural amplifier of Earth's energy, or something even more profound? Perhaps the answers lie not in proving one theory over another but in the experience itself—standing beneath the vast star-streaked sky, feeling the hum of the land, and embracing the unknown.

🌐 *Geophysical Note*

The San Luis Valley lies atop the Rio Grande Rift—an active fault system with geothermal flow, deep aquifers, and magnetic variation. These conductive layers may concentrate telluric currents and amplify subtle Earth energies.

For those drawn to the mysteries of the cosmos and the hidden forces of the Earth, the San Luis Valley stands as an open invitation. It is a place where the seen and unseen converge, where energies beyond explanation shape the land, and where those who listen may hear whispers of something greater than themselves.

NORTH AND SOUTH TABLE MOUNTAIN
VOLCANIC PLATEAUS OF ENERGY AND MYSTERY

Towering over Golden, Colorado, North and South Table Mountain are more than just scenic mesas—they are remnants of ancient volcanic activity that hold immense geological energy. These striking plateaus, with their distinctive flat tops and steep cliffs, are not only a geological wonder but also an electrically active location where natural forces converge in a dynamic interplay of magnetic fields, conductivity discontinuities, and atmospheric electricity.

◆ ENERGY SIGNATURE GUIDE ◆

Location: North & South Table Mountain, Golden, CO

Peak Energy Time: Midday, Solstices, and post-lightning storms

Energy Type: Basalt flows over sedimentary clays, magnetic contrasts, fossilized energy layers, and residual volcanic charge

Best Activity: Solar activation meditation, movement practices, and energy layering visualization

 Symbol: Flat-Topped Mesa – Represents stored ancient energy and stability

Formed by lava flows nearly 65 million years ago, these basalt-capped mesas have unique properties that make them prime locations for experiencing Earth's subtle electrical energy. The dense volcanic rock retains heat, interacts with atmospheric currents, and enhances telluric energy flow. The region's unique topography, combined with its elevated position above the surrounding terrain, makes it a hotspot for electromagnetic activity. It is no coincidence that visitors often report an uncanny sense of clarity, physical invigoration, and even heightened intuition when exploring these geological formations.

The mesas sit atop significant conductivity discontinuities, where different rock layers and underground mineral compositions create shifts in electrical resistance. These conditions foster the movement of natural electric currents known as telluric currents, which can subtly affect human physiology and perception. Some researchers suggest that locations with strong telluric activity may amplify the body's bioelectrical signals, leading to enhanced meditation experiences, sharper mental focus, and a profound sense of connection to the environment.

Magnetometers have detected localized variations in the Earth's magnetic field on and around the mesas, suggesting that subsurface geological features contribute to electromagnetic anomalies. Basalt, the dominant rock type of North and South Table Mountain, contains iron-rich minerals that interact with the planet's magnetic field. These interactions may create a subtle yet perceptible energy signature, drawing those sensitive to geomagnetic fluctuations. Ancient civilizations and indigenous tribes often sought such places for vision quests, ceremonies, and spiritual renewal.

Another fascinating aspect of these mesas is their role in atmospheric electricity. Due to their height, flat surfaces, and geological composition, North and South Table Mountain act as natural conductors of atmospheric charge, particularly during storms. This makes them prime locations for observing lightning activity and other electrical phenomena. Some visitors have described experiencing tingling sensations in their hands or a lightheaded feeling while standing atop the plateaus, possibly due to changes in the local electric field.

Today, the mesas are popular hiking and exploration destinations, attracting those who seek adventure and those who wish to immerse themselves in the natural energies of the land. Whether you are drawn to the towering rock formations for their geological intrigue, electromagnetic mysteries, or spiritual allure, North and South Table Mountain remain places where nature's pulse is palpable. They serve as reminders that the Earth is alive with energy, constantly shifting, resonating, and interacting with those who take the time to listen.

RED ROCKS PARK AND AMPHITHEATRE
A NATURAL CONDUCTOR OF ENERGY AND SOUND

Red Rocks Park and Amphitheatre is not just one of the most famous concert venues in the world—it is also a powerful energy center. Nestled between towering sandstone formations, the amphitheater's unique geological composition amplifies sound vibrations and electromagnetic activity. Many visitors report feeling a heightened sense of awareness, as though the land itself is alive with energy. The natural curvature of the rock formations, combined with the presence of iron-rich minerals, may contribute to the intense energetic presence felt here.

◆ ENERGY SIGNATURE GUIDE ◆

- **Location:** Red Rocks Park and Amphitheatre, Morrison, CO
- **Peak Energy Time:** Sunrise concerts, thunderstorms, and Summer Solstice
- **Energy Type:** Tilted Fountain Formation sandstone, acoustic resonance, iron oxide charge, and atmospheric amplification
- **Best Activity:** Sonic meditation, breathwork, and collective energy rituals
- **Symbol:** 🎵 Echo Spire – Represents soundwave alignment and energetic amplification

Geological Properties and Electromagnetic Activity

The rock formations at Red Rocks are composed primarily of ancient sandstone, dating back over 300 million years. These formations, rich in iron and quartz, contribute to the amphitheater's unique energetic frequency. Quartz, known for its piezoelectric properties, can generate electrical charges under mechanical stress, potentially influencing the electromagnetic field of the area. Additionally, the high iron content in the rocks may enhance conductivity, affecting the way energy flows through the land.

Many visitors, particularly musicians and spiritual seekers, describe an almost electric quality in the air, as though the site itself resonates with energy. Some speculate that Red Rocks sits on a natural ley line—an invisible energy grid believed to connect powerful places on Earth. Whether scientific or spiritual, the energetic intensity of the location is undeniable.

A Space for Sound and Spirituality

Red Rocks has been a gathering place for centuries. Indigenous tribes, including the Ute and Cheyenne, considered the area sacred and used it for ceremonies long before modern civilization recognized its unique acoustics. The amphitheater's ability to carry sound with perfect clarity creates an immersive experience that extends beyond music. Even during quiet moments, the rocks seem to hum, reflecting and amplifying subtle natural frequencies.

Meditation and energy work at Red Rocks can be particularly potent due to the site's natural amplification of sound waves and vibrational energy. Practitioners of yoga and breathwork find that their sessions take on a new depth in the amphitheater, where sound and silence interact in perfect harmony. The landscape itself seems to encourage reflection, connection, and heightened awareness.

The Connection Between Music and Energy

Music, an expression of frequency and vibration, finds its perfect stage at Red Rocks. The amphitheater has hosted legendary performances by artists who often describe a unique, almost mystical connection to the space. The natural acoustics, combined with the amphitheater's geological and energetic properties, make each performance an immersive experience not just for the audience but also for the performers themselves.

Concertgoers frequently report feeling an overwhelming sense of unity, euphoria, and spiritual connection during live performances. Some even believe that the music leaves an imprint on the land, continuously feeding the amphitheater's energy, making it an ever-evolving power spot.

Preserving the Energy of Red Rocks

While Red Rocks remains a beloved destination, its increasing popularity also brings concerns about preservation. Large crowds, infrastructure expansion, and environmental changes could disrupt the natural balance of the space. Conscious efforts to maintain the amphitheater's integrity—both physically and energetically—are essential to ensuring that future generations can experience its magic.

> ⚠ *Geological Tip*
> *The tilted Fountain Formation, rich in iron oxide, may help **direct energy upward**, while the exposed layers enhance grounding and vertical flow.*

For those seeking to explore Red Rocks beyond its musical legacy, a mindful approach is key. Whether attending a concert, hiking its scenic trails, or simply sitting in stillness among the towering formations, approaching the space with reverence can deepen one's connection to its energy. In many ways, Red Rocks serves as a reminder that music, nature, and human consciousness are intricately woven into the energy of the land.

THE MAROON BELLS

ALPINE MAJESTY AND THE RESONANCE OF QUARTZ

The Maroon Bells, often considered some of the most photographed peaks in the world, are more than just a scenic marvel. These mountains contain high concentrations of quartz, a mineral known for its piezoelectric properties, which generate subtle electrical charges when compressed. This may explain why many visitors experience a sense of calm and heightened intuition when hiking in the area. Additionally, the pristine alpine lakes and streams that surround the peaks further enhance the flow of natural energy.

◆ **ENERGY SIGNATURE GUIDE** ◆

- 📍 **Location:** Maroon Bells, near Aspen, CO
- ⏳ **Peak Energy Time:** Sunrise, Full Moon, and Late Summer
- ⚡ **Energy Type:** Layered mudstone and quartzite, glacial-carved basin, and reflective water-channeling resonance
- ♟ **Best Activity:** Mirror-gazing meditation, gratitude journaling, and stillness practices
- **Symbol:** Twin Peaks Reflection – Represents duality, inner clarity, and emotional echo

Geological Composition and Energy Properties

Rising over 14,000 feet above sea level, the Maroon Bells are composed of a unique combination of metamorphic rock and quartzite. The high quartz content of the mountains gives them a distinct vibrational frequency, making them an ideal location for energy-sensitive individuals. Quartz has been used for centuries in spiritual and metaphysical practices due to its ability to amplify energy, enhance clarity, and facilitate communication with higher consciousness. The natural compression of these quartz-rich formations over millions of years has resulted in an environment that subtly radiates energy, influencing those who visit.

The region's location within the Elk Mountains, combined with its high-altitude setting, means that electromagnetic energy circulates differently here compared to lower elevations. Some hikers and meditators report a tingling sensation, a sense of euphoria, or an enhanced ability to focus when spending time near the peaks. This could be attributed to the way quartz interacts with the Earth's natural electromagnetic field, creating a resonance that affects human physiology and perception.

Water as an Energy Conductor

The Maroon Bells are surrounded by crystal-clear alpine lakes, including Maroon Lake and Crater Lake, both of which reflect the grandeur of the peaks and play a significant role in the area's energetic flow. Water is a well-known conductor of energy, and its movement through quartz-rich terrain enhances the natural vibrational qualities of the region. Visitors often describe feeling an overwhelming sense of peace when gazing into the still waters, as if the lakes themselves act as mirrors of the Earth's energy.

Flowing streams and waterfalls originating from the high-altitude snowmelt further contribute to the dynamic energy of the area. The negative ions generated by these moving waters have been shown to elevate mood, reduce stress, and improve overall well-being. This may explain why people who visit the Maroon Bells leave feeling recharged and mentally refreshed.

Spiritual and Cultural Significance

Long before modern visitors discovered the Maroon Bells, Indigenous tribes recognized the region as a sacred place. The Ute people, who inhabited the surrounding valleys, considered the mountains to be the home of powerful spirits. They performed ceremonies and vision quests in the area, believing that the quartz formations held the wisdom of the Earth and could facilitate deep spiritual insights.

Even today, the Maroon Bells remain a pilgrimage site for seekers of natural wisdom. Many visitors come not only for the breathtaking scenery but also to connect with the land's profound energy. Meditation, grounding exercises, and quiet contemplation are common practices among those who wish to attune themselves to the high-frequency vibrations of the mountains.

Preserving the Energy of the Maroon Bells

As one of Colorado's most beloved natural landmarks, the Maroon Bells face challenges due to increasing tourism. With thousands of visitors each year, the delicate ecosystem is at risk of being disrupted. To maintain the area's pristine condition and energetic integrity, conservation efforts are essential. The U.S. Forest Service has implemented permit systems, shuttle services, and visitor guidelines to minimize human impact while allowing people to experience the magic of the landscape.

For those looking to connect deeply with the energy of the Maroon Bells, visiting during the quieter hours of dawn or dusk can provide a more immersive experience. Mindfulness, respect for the land, and a commitment to leaving no trace ensure that this sacred space continues to radiate its unique vibrational energy for generations to come.

THE BLACK CANYON OF THE GUNNISON
A GEOLOGICAL GATEWAY TO DEEP ENERGY

Carved over millions of years by the Gunnison River, the Black Canyon is a place where the Earth's deep energy is exposed in dramatic fashion. The canyon walls are composed of some of the oldest rock formations in North America, dating back nearly two billion years. These ancient structures act as conduits for telluric currents, and many visitors report feeling an overwhelming sense of reverence and connection to the land. The extreme depth and narrow width of the canyon create an amplified energetic field, making it a prime location for meditation and reflection.

◆ **ENERGY SIGNATURE GUIDE** ◆

Location: Black Canyon of the Gunnison, Western Colorado

Peak Energy Time: Noon shadows, Winter Solstice, and after rainstorms

Energy Type: Sheer crystalline cliffs of gneiss and schist, magnetic gradients, and echoing vertical current

Best Activity: Shadow meditation, echo-listening, and grounding breathwork

Symbol: ⬬ Deep Chasm – Represents descent, depth perception, and hidden power

Geological Composition and Energy Flow

The Black Canyon's sheer cliffs are composed primarily of Precambrian gneiss and schist, some of the most ancient rock formations on the continent. These rocks have endured immense pressure and heat, crystallizing into formations that conduct and store energy. Many geologists and spiritual seekers believe that such ancient landscapes hold deep Earth memory, carrying the energetic imprint of the planet's geological evolution.

The presence of telluric currents—naturally occurring electrical currents that flow through the Earth's crust—enhances the canyon's energetic profile. These currents interact with the canyon's rock formations, creating a field of subtle electromagnetic activity. Visitors who are sensitive to energy often describe tingling sensations, a sense of clarity, or even an emotional release when spending time within the depths of the canyon.

A Natural Vortex for Meditation and Reflection

The canyon's extreme depth, in some places plunging more than 2,700 feet, contributes to a feeling of vastness and introspection. Unlike the wide expanses of the Grand Canyon, the Black Canyon's narrow chasm creates an intense sense of enclosure, as if the Earth itself is wrapping around those who enter. This effect amplifies stillness and fosters a deep, meditative state.

Many spiritual practitioners find the Black Canyon to be an ideal location for grounding exercises, breathwork, and silent contemplation. The way sound is absorbed by the rock walls creates an otherworldly quiet, making it a perfect place to disconnect from distractions and tune into the subtle energies of the land. Whether sitting at the canyon's rim or hiking into its depths, the experience is often described as transformative.

Sacred and Cultural Significance

Long before modern visitors discovered its power, the Black Canyon was revered by Indigenous tribes. The Ute people, who lived in the region for centuries, considered the canyon a place of deep spiritual significance. They saw it as a portal to the underworld, where ancestors and spirits resided. The shadows that fill the canyon for most of the day reinforced the belief that it was a realm between worlds.

Ceremonies, vision quests, and sacred gatherings took place along the canyon's edges, where tribal members sought guidance from the spirits of the land. Even today, the canyon retains a sense of mystery, as if the whispers of the past still echo through its walls.

Preserving the Energy of the Black Canyon

As a designated national park, the Black Canyon of the Gunnison remains largely untouched by modern development. However, the increasing number of visitors raises concerns about preserving its pristine environment and energetic integrity. Staying on designated trails, minimizing noise pollution, and approaching the site with respect ensure that future generations can continue to experience its power.

For those seeking a profound connection to the Earth's oldest energies, the Black Canyon stands as a timeless sanctuary—an ancient chasm where geological history and spiritual awareness converge.

SHIFTING LANDSCAPES AND THE POWER OF THE EARTH

The Great Sand Dunes of Colorado are unlike any other landscape in the world. Their constant movement and shifting patterns generate unique electromagnetic properties. Some studies suggest that the friction created by the dunes' movement generates a low-frequency hum, which may be perceptible to those who are highly sensitive to energy. Indigenous tribes have long considered the dunes a place of spiritual significance, using them for vision quests and ceremonies. Today, visitors often report feeling deeply grounded and rejuvenated after spending time in this ever-changing environment.

◆ ENERGY SIGNATURE GUIDE ◆

Location: Great Sand Dunes National Park, San Luis Valley, CO

Peak Energy Time: Dawn, New Moon, and during seasonal winds

Energy Type: Quartz-rich sand, subsurface water flow, eolian charge buildup, and magnetic field interplay

Best Activity: Barefoot grounding walks, breath-focused meditation, and dream seeding under the stars

Symbol: Ripple Dune – Represents flow, impermanence, and subtle movement of intention

A Landscape in Motion

Rising up to 750 feet, the Great Sand Dunes are the tallest dunes in North America. They were formed by a unique combination of wind patterns, water flow, and geological forces. The perpetual movement of sand grains across the dunes creates an ever-evolving environment, symbolizing change and transformation. The shifting nature of this landscape is thought to contribute to its powerful energetic properties. Some visitors describe the sensation of walking on the dunes as grounding yet strangely electric, as though the Earth itself is alive beneath their feet.

Scientists have noted that the dunes exhibit a rare phenomenon known as "singing sands." When conditions are right, the movement of sand creates a deep, resonant hum, sometimes audible over great distances. This natural vibration aligns with theories that certain landscapes hold energetic frequencies that can influence human perception and emotional states.

Spiritual and Cultural Significance

For centuries, Indigenous tribes, including the Ute and Navajo, have revered the Great Sand Dunes as a place of great power. Many believe the dunes were created by spirits or as a sacred passage between worlds. Vision quests and spiritual ceremonies were often conducted in the dunes, as the landscape's constantly shifting form was thought to reflect the fluid nature of existence. The sand itself was believed to hold the prayers and wisdom of those who walked upon it.

The dunes also hold significance in oral traditions, with some stories speaking of guardian spirits that watch over the land. Even today, modern spiritual practitioners visit the dunes to engage in meditation, energy work, and rituals designed to enhance inner clarity and transformation.

Electromagnetic Properties and Energy Work

The combination of high-altitude wind activity, mineral-rich sands, and friction between grains creates an environment where subtle electromagnetic currents may be present. Energy healers and practitioners of Earth-based spirituality believe these conditions amplify natural energetic flows, making the dunes a prime location for grounding and attuning to Earth's frequencies.

⚡ *Electrostatic Insight*

Shifting grains in the dunes generate electrostatic charge, especially in dry, windy conditions. This frictional energy may subtly stimulate the body's own bioelectric field—especially through barefoot contact.

Visitors who engage in deep meditation or barefoot walking often describe feeling a noticeable shift in energy. Some report experiencing heightened intuition, vivid dreams, or an increased sense of connection to the land. This has led some researchers to speculate that the Great Sand Dunes function as a natural energy vortex, capable of enhancing spiritual experiences.

Preserving the Magic of the Dunes

As part of Great Sand Dunes National Park and Preserve, the dunes are protected from excessive development. However, climate change, increased tourism, and shifting water patterns continue to impact the delicate balance of this ecosystem. Practicing mindful exploration—such as staying on designated paths, respecting Indigenous traditions, and leaving no trace—helps ensure that the dunes retain their mystical energy for generations to come.

For those seeking transformation and connection with Earth's elemental forces, the Great Sand Dunes offer an unparalleled experience. In their shifting patterns, they remind us of the impermanence of all things, while their grounding presence invites us to reconnect with the deep, powerful energy of the land.

HIGH-ALTITUDE ENERGY VORTEX

Rocky Mountain National Park is not only a sanctuary for wildlife and breathtaking views but also a hub of intense natural energy. The combination of high-altitude winds, electromagnetic anomalies, and an abundance of quartz and granite formations creates an atmosphere charged with vibrational energy. Some specific locations, such as Dream Lake and Longs Peak, are considered energetic hotspots where visitors frequently experience heightened awareness, clarity, and even mystical encounters. The park's connection to celestial movements further enhances its significance, making it a must-visit for those seeking to tap into Colorado's natural power grid.

◆ ENERGY SIGNATURE GUIDE ◆

- **Location:** Rocky Mountain National Park, Northern Colorado
- **Peak Energy Time:** Sunrise at elevation, Summer Solstice, and during alpine storms
- **Energy Type:** Ancient granite uplift, glacial-carved peaks, quartz veins, and high-altitude atmospheric charge
- **Best Activity:** Breath-alignment hikes, sky-gazing meditation, and clarity rituals near alpine lakes
- **Symbol:** 🏔 Crowned Summit – Represents elevation of thought, stillness, and clear connection to higher states

Geological Composition and Energy Conduction

The park's foundation is composed primarily of ancient granite and quartz-rich formations. Quartz, known for its piezoelectric properties, generates an electrical charge under pressure, amplifying energy in the surrounding environment. This phenomenon, combined with the high-altitude air currents and changing atmospheric pressures, creates an ideal setting for natural energy vortices. Many visitors report feeling a tingling sensation, heightened mental clarity, or a profound sense of peace when spending time near these quartz-heavy rock formations.

Additionally, the park's rugged peaks and deep valleys influence the way electromagnetic energy moves through the landscape. Some areas, particularly those with significant rock formations, act as natural amplifiers, intensifying the energy flow and creating an optimal environment for meditation and spiritual reflection.

Energetic Hotspots within the Park

Several locations in Rocky Mountain National Park are particularly known for their energetic properties. These include:

Dream Lake: Often cited as one of the most mystical locations in the park, Dream Lake has a glass-like surface that reflects both the grandeur of the mountains and the subtle energies flowing through the landscape. Many visitors report experiencing deep relaxation and vivid dreams after spending time here.

Longs Peak: As one of Colorado's iconic 14ers, Longs Peak is known for its powerful presence. Hikers who ascend the peak often describe an intense surge of energy as they near the summit, as though stepping into a higher frequency.

The Alpine Tundra: The high-altitude tundra of Rocky Mountain National Park is home to some of the most extreme weather patterns and energetic phenomena. The thin air, strong winds, and wide-open landscapes create an expansive field of energy that enhances sensory awareness and deepens the connection to the natural world.

Celestial Alignments and Cosmic Energy

Rocky Mountain National Park is also uniquely tied to celestial movements. The alignment of certain peaks with the solstices and equinoxes suggests that ancient cultures may have recognized this land as a sacred observatory. Even today, stargazers and spiritual seekers flock to the park to witness meteor showers, eclipses, and planetary alignments, all of which seem to amplify the park's already potent energy.

Preserving the Energy of the Park

As more people seek to experience the energetic wonders of Rocky Mountain National Park, conservation efforts become increasingly important. Practicing responsible tourism—staying on trails, respecting wildlife, and minimizing environmental impact—ensures that this sacred energy vortex remains intact for future generations.

CRIPPLE CREEK

GOLD, QUARTZ, AND THE SPIRIT OF THE EARTH

Cripple Creek is one of Colorado's most historically significant mining towns, but beyond its gold rush legacy, it is an area teeming with electromagnetic anomalies and natural energy currents. The high concentration of quartz in the area, combined with deep underground mineral deposits, creates a geoelectric environment that amplifies energetic sensations. Many visitors report an almost electric charge in the air, especially in abandoned mines where raw minerals remain untouched.

◆ ENERGY SIGNATURE GUIDE ◆

📍 **Location:** Cripple Creek, Colorado

⌛ **Peak Energy Time:** High noon, Autumn Equinox, and during sudden weather shifts

⚡ **Energy Type:** Volcanic caldera remnants, gold-bearing quartz veins, mineralized faults, and subsurface conductivity

🧘 **Best Activity:** Solar-based grounding, prosperity-focused meditation, and energetic excavation journaling

🎯 **Symbol:** 💰 Golden Vein – Represents inner excavation, abundance consciousness, and deep energetic potential

A Geological Powerhouse

Nestled within the Pikes Peak granite formation, Cripple Creek is geologically unique. The region is rich in quartz, feldspar, and other crystalline minerals that naturally conduct and amplify energy. Quartz, in particular, is known for its piezoelectric properties—when subjected to pressure, it generates an electrical charge. This characteristic may explain why many visitors experience heightened awareness, increased intuition, and even subtle vibrational sensations when exploring the area.

Additionally, the town sits on a network of deep underground veins rich in gold, silver, and tellurium. These minerals not only played a significant role in the town's historic gold rush but also contribute to its energetic intensity. The presence of conductive metals beneath the surface creates a natural circuit, allowing electromagnetic energy to flow through the land in ways that are often felt rather than seen.

Mystical Sensations and Energetic Anomalies

Beyond its geological composition, Cripple Creek is known for its unusual energy anomalies. Many visitors and paranormal investigators report experiencing unexplained sensations, such as tingling skin, lightheadedness, or a buzzing sensation when standing near certain rock formations or abandoned mining shafts. Some even claim to hear low-frequency hums emanating from deep within the earth, possibly due to subterranean water movement or mineral interactions beneath the surface.

Several of the region's abandoned mines, including the historic Mollie Kathleen Mine, have gained reputations as locations where people frequently feel the presence of unseen forces. Whether this is due to residual energy left from the intense labor and hardship of miners or an innate geophysical phenomenon remains open to interpretation. Regardless, there is no denying that the area holds a mysterious and almost magnetic pull.

Cripple Creek's Spiritual and Paranormal Connections

Cripple Creek's unusual energetic properties have drawn interest from both spiritual seekers and paranormal enthusiasts. The town is home to numerous ghost stories, many of which center around its historic buildings and abandoned mines. Some believe that the high mineral content in the area acts as an energy conduit, preserving emotional imprints from the past and making them more perceptible to sensitive individuals.

For those seeking a more meditative experience, Cripple Creek offers numerous locations ideal for energy work and deep reflection. The natural quartz formations scattered throughout the area provide powerful grounding energy, while the remote mountainous terrain fosters a sense of tranquility and connection to the earth.

Preserving the Energy of Cripple Creek

While Cripple Creek's mining days may be behind it, the energy of the land remains. As interest in energy hotspots grows, so does the need for conservation and respectful exploration. Visitors can honor the spirit of the area by practicing mindfulness, minimizing environmental impact, and approaching the land with gratitude and reverence. Whether exploring historic sites, meditating in quartz-laden fields, or simply absorbing the mysterious energy of the region, Cripple Creek remains a fascinating and powerful destination for those drawn to the spirit of the earth.

PIKES PEAK
A TOWERING BEACON OF ELECTROMAGNETIC ENERGY

Pikes Peak is not just a stunning landmark—it is also an energy vortex. Rising above the landscape at 14,115 feet, this mountain interacts directly with atmospheric energy and electromagnetic activity. The granite-rich terrain and exposure to high-altitude winds create an energetic atmosphere that has drawn seekers, spiritualists, and adventurers alike. Many experience a sense of heightened awareness, clarity, and even altered states of consciousness when ascending the peak.

◆ ENERGY SIGNATURE GUIDE ◆

- **Location:** Pikes Peak, Colorado Springs, CO
- **Peak Energy Time:** Sunrise above the clouds, Equinoxes, and during lightning-filled storms
- **Energy Type:** Pikes Peak Granite with massive quartz and feldspar crystals, piezoelectric potential, high-altitude ionization, and sky-earth current exchange
- **Best Activity:** Summit meditation, breath expansion, and energy anchoring at elevation
- **Symbol:** Crown Crystal – Represents vertical connection, visionary insight, and amplified energetic clarity

Geological Composition and Energy Influence

Pikes Peak is composed primarily of Pikes Peak granite, an ancient rock formation rich in quartz and feldspar. Quartz is well known for its piezoelectric properties, meaning it generates an electrical charge when subjected to mechanical stress. Given the immense pressures and shifting tectonic forces that have shaped the peak over millions of years, it's no surprise that the mountain carries a palpable energetic signature.

Additionally, the presence of iron and other conductive minerals within the rock amplifies electromagnetic energy. Scientists have long studied the role of Earth's natural electromagnetic fields in affecting human physiology, and locations with high quartz content—such as Pikes Peak—are often reported as hotspots of energetic activity.

The High-Altitude Energy Effect

The extreme elevation of Pikes Peak plays a crucial role in its energetic impact. At over 14,000 feet, the air is thin, oxygen levels are reduced, and atmospheric pressure is lower. These conditions can create altered states of consciousness, often described as a feeling of euphoria or mental clarity. Many hikers and climbers report a deepened connection to the surrounding landscape, as if the mountain's energy amplifies their inner awareness.

Another key factor is the high-altitude winds that sweep across the summit. Wind itself carries energetic charge, and as it moves across the rocky surfaces of Pikes Peak, it is believed to generate and distribute subtle vibrational frequencies. This might explain why many people feel an electrifying sensation or tingling in their extremities when standing at the peak.

Spiritual and Cultural Significance

Long before modern visitors climbed to the summit, Indigenous tribes such as the Ute and Arapaho recognized Pikes Peak as a sacred place. They considered it a bridge between the Earth and the sky, a location where one could connect with spiritual forces. Vision quests and ceremonies were often held in the shadow of the peak, as tribal members sought guidance and wisdom from the spirits of the land.

Today, Pikes Peak continues to attract spiritual seekers, energy workers, and those drawn to the mountain's powerful presence. Some believe that ley lines—mystical alignments of Earth's energy—converge in this region, further enhancing the mountain's status as an energy vortex.

Preserving the Power of Pikes Peak

As a popular destination, Pikes Peak sees thousands of visitors each year. While its energy remains strong, increased human activity can have an impact on the mountain's delicate ecosystem. Conservation efforts, including minimizing pollution, respecting wildlife, and staying on designated trails, ensure that Pikes Peak's natural and energetic beauty remains intact for future generations.

Whether one seeks adventure, spiritual insight, or simply a moment of profound connection with nature, Pikes Peak stands as a towering beacon of electromagnetic energy—offering an experience that is both grounding and transcendent.

ANCIENT CIVILIZATION AND EARTH ENERGY

Mesa Verde is a sacred place where the ancient Ancestral Puebloans once thrived. Their cliff dwellings are positioned in alignment with celestial movements, reinforcing the idea that this location was chosen for its spiritual and energetic properties. The combination of sandstone formations, underground aquifers, and its rich cultural history makes this one of the most fascinating energy centers in Colorado.

◆ ENERGY SIGNATURE GUIDE ◆

Location: Mesa Verde, Southwestern Colorado

Peak Energy Time: Sunset, Winter Solstice, and moments of stillness within alcoves

Energy Type: Layered sandstone mesas, ancestral imprints, cardinal alignment, and canyon acoustics

Best Activity: Silent reflection in cliff dwellings, ancestral reconnection meditations, and ceremonial breathwork

Symbol: Sacred Hollow – Represents memory in stone, rooted wisdom, and the echo of ancient presence

Geological and Energetic Foundations

Mesa Verde, meaning "Green Table" in Spanish, is a massive plateau composed primarily of sandstone. This porous rock type interacts with underground water channels, creating a natural conductivity that enhances energy flow. Sandstone formations in energy-rich locations often act as natural amplifiers, absorbing and distributing electromagnetic frequencies through the land. Some visitors report feeling a distinct charge in the air, particularly around the dwellings and ancient kivas, spaces once used for sacred ceremonies.

Beneath the surface, underground aquifers flow beneath the plateau, further contributing to the energetic properties of the area. Water is known to enhance telluric currents, which are natural electrical currents running through the Earth's crust. The presence of these currents, combined with the high mineral content in the soil, creates a potent energetic field that may explain why this region was so significant to the Ancestral Puebloans.

Celestial Alignment and Sacred Geometry

The Ancestral Puebloans were master astronomers, carefully aligning their structures with celestial bodies. Many of the dwellings within Mesa Verde are oriented to track the solstices and equinoxes, suggesting an advanced understanding of Earth's energetic and cosmic rhythms. Sunlight entering key structures at specific times of the year highlights the intentionality behind their construction, reinforcing the belief that this was a place of deep spiritual connection.

In addition to celestial alignments, the layout of the structures within the park reflects sacred geometry principles. Kivas, circular subterranean rooms used for ceremonies, often feature key alignments that mirror constellations and the movement of the sun and moon. These architectural choices may have amplified the energetic properties of the land, creating spaces where ritual and meditation could be enhanced by the natural forces present.

Spiritual Legacy and Modern-Day Energetic Experiences

Mesa Verde remains a place of profound spiritual significance. Many visitors report a sense of peace, grounding, and connection when walking among the ancient ruins. Some believe that the wisdom and energy of the Ancestral Puebloans still linger in the stones, offering guidance to those who come with an open heart and mind. Sensitives and energy workers often describe feeling subtle vibrations, shifts in consciousness, or a heightened sense of intuition while meditating within the dwellings.

The region is also known for unexplained phenomena, including orbs of light, sudden changes in temperature, and an overwhelming sense of being "watched" by unseen forces. Whether these occurrences are spiritual imprints or the result of natural electromagnetic activity remains a topic of intrigue.

Preserving the Energy of Mesa Verde

As a UNESCO World Heritage Site, Mesa Verde is protected from excessive development, ensuring that its energy remains intact. Visitors are encouraged to respect the land by following conservation guidelines, staying on designated paths, and approaching the site with reverence. By honoring the wisdom of the past and preserving the integrity of the landscape, we ensure that future generations can continue to experience the magic of this ancient, energy-rich sanctuary.

CRESTONE

A SPIRITUAL AND ENERGETIC MECCA

Crestone is often referred to as the spiritual heart of Colorado. Nestled at the base of the Sangre de Cristo Mountains, this small town is home to an array of spiritual centers, including Buddhist monasteries, Hindu ashrams, and Native American ceremonial sites. The high vibrational energy of the land, combined with its geological features, makes Crestone a destination for those seeking deep spiritual connection and energetic renewal.

◆ ENERGY SIGNATURE GUIDE ◆

Location: Crestone, San Luis Valley, CO

Peak Energy Time: Equinox sunrises, midnight stillness, and during solar flares

Energy Type: Intersection of fault zones, aquifer flow, quartz-laced bedrock, and geomagnetic anomaly influence

Best Activity: Deep meditation, chakra alignment, and interspiritual ceremony

Symbol: ✦ Convergence Spiral – Represents spiritual synthesis, grounding through openness, and vertical inner attunement

---

Geological and Electromagnetic Properties

The geological foundation of Crestone plays a significant role in its energetic presence. The town is located within a region rich in quartz and other crystalline minerals, which have long been associated with energy amplification and spiritual attunement. Quartz, in particular, is known for its piezoelectric properties, meaning it generates electrical charges when under pressure. Some energy workers believe this property enhances the vibrational frequency of the land, making it easier to access states of deep meditation and heightened awareness.

Additionally, Crestone sits near the San Luis Valley, an area known for its unusual electromagnetic activity. The valley is home to numerous reports of unexplained lights, UFO sightings, and energetic phenomena. Some researchers speculate that underground water channels and fault lines contribute to these anomalies by generating subtle electrical currents that interact with the Earth's magnetic field. Whether viewed through a scientific or spiritual lens, Crestone's unique landscape undeniably contributes to its status as a powerful energetic hub.

A Haven for Spiritual Seekers

Crestone is home to one of the highest concentrations of spiritual centers in the United States, making it a global destination for seekers of enlightenment and inner peace. Buddhist stupas, Hindu temples, Zen meditation centers, and Native American ceremonial grounds all coexist in harmony, each offering a unique pathway to spiritual growth. Visitors can participate in silent retreats, prayer ceremonies, energy healing sessions, and meditative practices that draw upon the land's powerful energy.

One of the most well-known institutions in Crestone is the Crestone Mountain Zen Center, which offers deep immersion into Buddhist teachings in a serene, high-altitude setting. Nearby, the Haidakhandi Universal Ashram provides a space for Hindu devotional practices, including fire ceremonies and mantra chanting. Native American traditions are also honored in the region, with sacred sweat lodges and vision quests frequently conducted on the land.

Energetic Phenomena and Personal Experiences

Many visitors report profound energetic shifts upon arriving in Crestone. Some describe feeling a heightened sense of awareness or an immediate sense of calm and grounding. Others report experiencing vivid dreams, spontaneous spiritual insights, or even encounters with unexplained phenomena. The region's isolation and pristine natural beauty contribute to these experiences, allowing individuals to disconnect from the distractions of modern life and tune into the subtle energies of the land.

Preserving the Spiritual Integrity of Crestone

As Crestone gains popularity as a spiritual destination, it is important to protect the land's energy and sacred spaces. Conservation efforts ensure that the region remains undeveloped and pristine, allowing future generations to benefit from its powerful presence. Visitors are encouraged to approach Crestone with respect, honoring its traditions and the deep spiritual history embedded within the land. Whether one comes for a short retreat or a lifelong journey, Crestone offers a gateway to profound transformation and energetic renewal.

THE FLATIRONS

MAGNETIC FORCES IN BOULDER'S ICONIC FORMATIONS

The Flatirons in Boulder are more than just a stunning backdrop to the city—they are natural energy conduits. The iron-rich sandstone formations interact with the Earth's magnetic field, creating an unusual energetic environment. Many visitors report feeling a sense of grounding and stability while hiking among these formations, as if they are being recalibrated by the land itself.

◆ ENERGY SIGNATURE GUIDE ◆

Location: The Flatirons, Boulder, CO

Peak Energy Time: Sunrise, Autumn Equinox, and during wind shifts

Energy Type: Tilted quartzite slabs, tectonic uplift stress, piezoelectric charge, and wind-enhanced polarity flow

Best Activity: Standing meditations, heart-opening breathwork, and sunrise walks along the stone faces

Symbol: ⊠ Blade Stone – Represents focus, energetic clarity, and amplified inner signal

---

Geological Significance and Magnetic Properties

The Flatirons, composed of conglomerate sandstone from the Fountain Formation, date back nearly 300 million years. These massive rock slabs tilt at striking angles, forming a dramatic gateway to the Rocky Mountains. The presence of iron and other magnetic minerals within the rock enhances their conductivity, allowing the formations to subtly interact with geomagnetic fields.

Studies have shown that iron-rich geological formations can produce localized magnetic anomalies. Visitors who are sensitive to energy often describe an almost imperceptible pull when walking near these formations, as if the land itself is adjusting their internal compass. Some hikers have even reported their compasses behaving erratically in certain areas, which supports the idea that these formations may have a measurable magnetic influence.

Energetic Effects on Visitors

The Flatirons are often cited as a place where people feel more balanced, focused, and physically energized. Some believe that the iron within the rocks helps stabilize the body's bioelectric field, which can be disrupted by modern electromagnetic pollution. Others describe an enhanced connection to the Earth, particularly when sitting directly on the rocks during meditation or mindfulness practices.

The sloping angles of the formations also contribute to their energetic properties. Some esoteric traditions suggest that natural pyramidal or slanted shapes, like those found in sacred architecture, can amplify energy flow. Whether or not this effect is scientifically verifiable, those who spend time among the Flatirons frequently report a heightened sense of well-being and mental clarity.

Historical and Spiritual Significance

Long before the city of Boulder was founded, the Indigenous Ute and Arapaho tribes considered the Flatirons a sacred landscape. These towering rock formations were seen as a bridge between the physical and spiritual worlds, a place where one could commune with nature and receive guidance from the ancestors. Many Indigenous ceremonies were conducted in the foothills surrounding the Flatirons, with the belief that the land held special wisdom and healing energy.

Today, the Flatirons continue to draw spiritual seekers, healers, and energy workers who wish to experience the site's vibrational frequencies. Yoga practitioners and meditation groups frequently gather in the meadows below the formations, using the site's natural serenity to deepen their practice.

Conservation and Ethical Visitation

As one of Boulder's most popular hiking and climbing destinations, the Flatirons face increasing foot traffic and environmental impact. Visitors are encouraged to stay on designated trails to preserve the delicate vegetation and respect the natural energy of the land. Practicing mindfulness, leaving no trace, and engaging with the site in a respectful manner ensures that the Flatirons remain an energetic sanctuary for generations to come.

Whether one visits for adventure, spiritual renewal, or scientific curiosity, the Flatirons stand as an enduring testament to the powerful forces shaping both the Earth and human consciousness.

THE ANIMAS RIVER VALLEY
FLOWING CURRENTS OF ENERGY AND HISTORY

The Animas River, which flows through Durango and beyond, is a vital waterway that has long been recognized for its powerful energetic presence. The movement of water, especially through areas rich in minerals, creates an electric charge that many believe enhances intuitive perception and spiritual clarity.

◆ ENERGY SIGNATURE GUIDE ◆

📍 **Location:** Animas River Valley, near Durango, CO

⏳ **Peak Energy Time:** Morning mist, Spring Equinox, and during snowmelt runoff

⚡ **Energy Type:** Flowing groundwater currents, mineral-rich sediment layers, magnetic contrasts from nearby igneous intrusions, and rhythmic pulse from water movement

🧘 **Best Activity:** Flow-state walking meditation, water invocation rituals, and gratitude practice

🎯 **Symbol:** Serpent Stream – Represents life force in motion, cleansing, and intuitive momentum

---

Geological Influence and Energetic Properties

The Animas River originates in the rugged San Juan Mountains and carves its way through some of the most mineral-rich terrain in the Southwest. The riverbed is lined with deposits of gold, silver, quartz, and other conductive minerals, contributing to its unique electromagnetic properties. The friction created by the river's constant movement over these mineral deposits generates subtle energy fields that may influence the human bioelectric system.

Many visitors to the Animas River Valley report heightened intuition, vivid dreams, and an increased sense of presence when near the river. The phenomenon of water amplifying energy is well-documented, with moving bodies of water known to produce negative ions—molecules that enhance mood, mental clarity, and physical well-being. This could explain why time spent near the Animas River often results in a deep sense of renewal and connection to nature.

A Sacred and Historical Waterway

Long before modern settlements, Indigenous tribes such as the Ute and Navajo revered the Animas River as a sacred waterway. They believed that the river carried spiritual messages and that its waters had the power to cleanse the soul. Ceremonial rituals, including water blessings and offerings, were performed along its banks, reinforcing the belief that the river was a living force with its own consciousness.

Spanish explorers named the river "Rio de las Animas Perdidas," or "River of Lost Souls," a name steeped in legend. Some say it was given after a group of explorers perished in its turbulent waters, while others claim it refers to the river's mysterious, otherworldly quality. Today, the river retains its enigmatic energy, drawing spiritual seekers and energy workers who wish to tap into its vibrational frequency.

Energetic Effects on Visitors

The Animas River Valley serves as a natural energy conduit, with its continuous flow generating a dynamic field of movement and transformation. Some individuals claim to experience enhanced meditation, spontaneous insights, or even visions when sitting beside the river or dipping their hands into its waters. Others find that the rhythmic sound of rushing water induces a trance-like state, promoting deep relaxation and mental clarity.

Energy healers and shamanic practitioners often use the river's energy in their work, believing it to be a powerful ally in spiritual purification. Water, after all, has long been associated with emotional healing, adaptability, and the subconscious mind.

Preserving the Energy of the Animas River

Despite its beauty and power, the Animas River has faced environmental threats, including pollution from mining runoff and industrial activities. Conservation efforts are crucial in maintaining the purity of its waters and ensuring that future generations can continue to experience its energetic and spiritual benefits.

By approaching the Animas River with reverence, respect, and ecological mindfulness, visitors can preserve not only its physical integrity but also its profound energetic essence, keeping it a vital source of inspiration and renewal for years to come.

DEEP CANYONS AND CONCENTRATED ELECTROMAGNETIC FIELDS

The Royal Gorge, often referred to as the "Grand Canyon of the Arkansas," is a dramatic chasm carved by the Arkansas River over millions of years. The immense depth of the gorge, coupled with its composition of iron-rich granite and metamorphic rock, makes it a natural conduit for electromagnetic energy. The canyon walls act as amplifiers, channeling energy in ways that many visitors can physically sense.

◆ ENERGY SIGNATURE GUIDE ◆

Location: Royal Gorge, near Cañon City, CO

Peak Energy Time: Midday sun, post-thunderstorm, and early spring thaw

Energy Type: Vertical granite canyon walls, concentrated river energy, piezoelectric quartz seams, and air ionization from water-and-rock interaction

Best Activity: Solar stillness, sound-echo meditation, and energetic anchoring at canyon rim

Symbol: Stone Gate – Represents inner resilience, pressure-born insight, and high-current awareness

Geological Composition and Electromagnetic Activity

The Royal Gorge is composed primarily of Precambrian granite, gneiss, and schist—some of the oldest rock formations in North America. These dense rock types contain high concentrations of iron and quartz, both of which play significant roles in conductivity and energy transmission. Quartz, known for its piezoelectric properties, generates electrical charges when subjected to pressure, potentially intensifying the region's energetic field. The presence of iron-rich minerals also enhances the area's natural magnetism, contributing to the sensation of heightened awareness reported by many visitors.

The Arkansas River, which carved the gorge, continuously flows through its depths, creating an interaction between water and rock that further amplifies its electromagnetic properties. Water moving through mineral-rich terrain can generate subtle electric fields, adding another layer to the energetic intensity of the gorge.

Wind Patterns and Atmospheric Resonance

The towering cliffs of the Royal Gorge create unique wind patterns that add another dimension to its energetic environment. Wind moving through the narrow canyon accelerates as it is funneled between rock walls, producing vibrations and sound frequencies that many people describe as calming, almost meditative. Others report feeling an electric charge in the air, as if the natural elements combine to create an energetic vortex.

Scientific studies suggest that specific wind patterns can generate infrasound—low-frequency sound waves that are below the threshold of human hearing but can still influence the body's nervous system. These waves may explain the sense of deep relaxation or heightened alertness experienced by those who spend time in the gorge. Many visitors describe feeling as if they are being energetically "reset" while standing at its rim or hiking its trails.

Cultural and Spiritual Significance

Indigenous tribes, including the Ute and Apache, revered the Royal Gorge as a place of great power. Oral traditions describe the canyon as a portal between worlds, where energy from deep within the Earth meets the sky. Shamans and spiritual leaders conducted ceremonies at the gorge, using its natural energy to enhance visions, healing rituals, and communication with ancestors.

Even today, the Royal Gorge remains a site of spiritual pilgrimage. Some visitors engage in meditation or breathwork at key vantage points along the rim, seeking to harmonize their own energy with that of the land. Others report spontaneous moments of clarity or emotional release, as if the canyon's depth mirrors their own inner landscapes.

Preserving the Energetic Integrity of the Gorge

While the Royal Gorge is a popular tourist destination, it is crucial to approach it with respect. Visitors should stay on designated paths, minimize noise pollution, and refrain from disrupting the natural rock formations. By treating the land with reverence, we can ensure that its powerful energy remains intact for future generations to experience.

Whether one visits for adventure, reflection, or spiritual exploration, the Royal Gorge offers an unparalleled connection to the Earth's deep energy— reminding us of the immense forces that shape both landscapes and human consciousness.

THE SANGRE DE CRISTO MOUNTAINS
A SACRED AND MYSTERIOUS RANGE

The Sangre de Cristo Mountains are steeped in mystery, legend, and high vibrational energy. Spanning southern Colorado into New Mexico, these towering peaks hold a long history of spiritual significance for Native American tribes, including the Ute and Pueblo peoples. The name "Sangre de Cristo," meaning "Blood of Christ," was given by Spanish explorers who were struck by the intense red hues of the mountains at sunrise and sunset—a phenomenon that only adds to their mystical aura.

◆ ENERGY SIGNATURE GUIDE ◆

Location: Sangre de Cristo Range, Southern Colorado

Peak Energy Time: Twilight, Full Moon, and Solstice sunrises

Energy Type: Fault-driven uplift, quartz-rich granite peaks, deep aquifer channels, and sky-ground energetic exchange

Best Activity: Prayer hikes, moonlit meditation, and soul-alignment rituals at alpine thresholds

Symbol: Blood Sky Peak – Represents ascension, sacred tension, and ancestral convergence

Geological Composition and Energetic Influence

Geologically, the Sangre de Cristo Mountains are composed of quartzite, granite, and other crystalline minerals that naturally conduct and amplify electromagnetic energy. Quartz, known for its piezoelectric properties, generates electrical charges when under pressure, which may contribute to the unique vibrational presence of the range. The granite formations also enhance conductivity, creating an environment where electromagnetic fields interact with the land's geological structure.

Some researchers speculate that the unique mineral composition of the range interacts with telluric currents—natural electrical currents flowing through the Earth—to create an energetic field that enhances perception. Visitors often report heightened awareness, spontaneous insights, and a deepened sense of connection with nature. Some have even encountered unexplained lights in the region, leading to speculation that the mountains hold a portal-like quality.

Spiritual and Cultural Significance

The Indigenous peoples who inhabited these lands recognized the mountains as sacred, believing them to be home to spirits and powerful natural forces. Vision quests, ceremonial dances, and prayer rituals were commonly conducted in the remote high-altitude valleys, where the veil between the physical and spiritual worlds was believed to be thin.

The region is also known for its legends of hidden cities and lost civilizations. Stories passed down through generations speak of ancient beings and spiritual entities residing in the mountains, watching over those who seek enlightenment or transformation. Even today, modern-day spiritual seekers visit the range, drawn by its reputation as a site of high vibrational energy and inner awakening.

Crestone: A Spiritual Epicenter

The Sangre de Cristo Mountains are also home to Crestone, a town known for its diverse spiritual communities and energy vortexes. Nestled at the base of the range, Crestone is a hub for meditation centers, monasteries, and retreat spaces that attract seekers from around the world. It is said that ley lines—energetic pathways running through the Earth—converge in this region, further enhancing its metaphysical properties.

A Convergence of Spiritual Traditions

Crestone is unique in that it hosts an unusually high concentration of spiritual centers, representing a variety of traditions. Buddhist stupas, Hindu ashrams, Sufi meditation circles, Christian monasteries, and Native American ceremonial grounds all coexist harmoniously in this small town. The presence of these diverse traditions suggests that the land itself has a unifying and transformative quality, drawing people from all walks of life to explore their spiritual paths.

Many spiritual seekers believe that the Sangre de Cristo Mountains hold an ancient wisdom encoded in the land. The quiet isolation, pristine air, and stunning mountain views create an environment conducive to deep meditation and introspection. Visitors often report experiencing heightened awareness, profound insights, and even mystical visions while engaging in spiritual practices here.

Ley Lines and Energetic Vortexes

One of the most compelling aspects of Crestone is its connection to Earth's ley lines—subtle energy channels believed to flow across the planet. Some geomancers suggest that Crestone sits at a major intersection of these ley lines, which may explain the town's magnetic appeal to spiritual teachers, healers, and seekers. Energy vortexes, similar to those found in Sedona, Arizona, are said to be scattered throughout the area, acting as amplifiers of consciousness and transformation.

Hiking through the mountains or sitting in quiet meditation along the foothills, many visitors claim to feel a pulsing or tingling sensation, as if the land itself is transmitting energy. This phenomenon is often attributed to the quartz and crystalline minerals present in the surrounding rock formations, which are known for their ability to store and conduct energy.

Sacred Practices and Retreat Centers

Crestone offers numerous retreat opportunities for those looking to immerse themselves in spiritual practice. The Crestone Mountain Zen Center provides intensive meditation retreats, while the Haidakhandi Universal Ashram hosts fire ceremonies and devotional chanting. Additionally, many private retreat spaces offer solitude for contemplation and energy work.

Sweat lodges, vision quests, and shamanic ceremonies are also held in the area, honoring the region's Indigenous traditions. Respecting these practices and the land is an essential part of experiencing Crestone's energy in an authentic and meaningful way.

Honoring and Preserving the Energy of Crestone

As interest in Crestone's energetic and spiritual offerings grows, it is vital to ensure that the land remains undisturbed and sacred. Conservation efforts aim to protect the fragile alpine ecosystem while allowing visitors to experience the area's profound energy. Approaching Crestone with reverence, mindfulness, and a willingness to listen to the land ensures that this spiritual epicenter remains a sanctuary for transformation and connection for generations to come.

ANCIENT FOSSILS AND GEOLOGICAL RESONANCE

Dinosaur Ridge, located just outside of Denver, is an internationally recognized site for its prehistoric fossil beds and unique geological formations. Beyond its scientific significance, the ridge is also an energetic hotspot due to its composition of iron-rich rock, sandstone, and preserved dinosaur footprints that date back over 100 million years.

◆ ENERGY SIGNATURE GUIDE ◆

📍 **Location:** Dinosaur Ridge, Morrison, CO

⏳ **Peak Energy Time:** Midday sun, after rain, and near equinox alignment

⚡ **Energy Type:** Exposed Mesozoic layers, fossil-charged sandstone, magnetic mineral streaks, and residual imprint of ancient biological energy

🔱 **Best Activity:** Time-tracking meditation, grounding near fossil beds, and memory activation walks

🎯 **Symbol:** Stone Spiral – Represents deep time, ancestral memory, and the echo of primordial life

A Geological Time Capsule

The landscape of Dinosaur Ridge is a living record of Earth's deep history. The site preserves not only the footprints of massive dinosaurs but also the movements of ancient seas, shifting continents, and volcanic activity. Walking along the ridge, visitors can see fossilized impressions of prehistoric creatures, ripple marks from ancient shorelines, and layers of rock that reveal different geological eras. This tangible connection to the distant past creates an undeniable sense of wonder, as if the land itself whispers the stories of its ancient inhabitants.

Fossils, like quartz, are believed to store energy, carrying the imprint of Earth's deep past. Some energy workers suggest that fossilized remains act as conduits for ancient vibrations, holding onto the essence of a time when the planet's energy was vastly different. Many visitors report a sense of stepping back in time when walking the ridge, as if they can feel the ancient presence of the dinosaurs that once roamed the area. This phenomenon aligns with the idea that certain landscapes hold energetic imprints of the past, affecting those who are sensitive to subtle vibrations.

Electromagnetic and Magnetic Phenomena

One of the most intriguing aspects of Dinosaur Ridge is its proximity to the Dakota Hogback, a geologic uplift that forms a natural boundary between the Great Plains and the Rocky Mountains. This uplift contains a variety of magnetic minerals, including hematite and magnetite, which can influence compasses and electrical instruments. Scientists have documented minor geomagnetic anomalies in the area, suggesting that the presence of these minerals interacts with the Earth's electromagnetic field.

Some visitors believe this energy contributes to an enhanced sense of intuition and connection to Earth's primordial forces. Those who are energy-sensitive often describe a tingling sensation in their hands or feet while touching the fossilized footprints, as if they are tapping into an ancient frequency embedded in the rock. Meditation practitioners have noted that sitting atop the ridge allows them to feel grounded and deeply connected to Earth's history, as though they are receiving insights from a prehistoric past.

A Portal to Deep Time

For those seeking a unique energetic experience, exploring Dinosaur Ridge offers a chance to tap into the vast timelines of planetary history. Whether marveling at ancient footprints or meditating on the ridge's elevated plateaus, visitors often leave with a sense of awe at the enduring power and energy of the land itself. The landscape serves as both a scientific archive and an energetic gateway, offering a rare opportunity to connect with Earth's immense geological and biological history.

As visitation increases, it remains essential to respect the delicate fossils and geological features that make Dinosaur Ridge so special. By approaching the site with reverence and curiosity, we ensure that its energy and historical significance remain intact for generations to come.

DINOSAUR NATIONAL MONUMENT
VAULT OF THE ANCIENTS

Where the Green and Yampa Rivers carve their way through layered rock and time itself seems to ripple underfoot, Dinosaur National Monument pulses with a unique and ancient kind of power. Straddling the Colorado-Utah border, this site is best known for its rich fossil beds and exposed slices of Mesozoic history—but its energetic resonance runs much deeper.

◆ ENERGY SIGNATURE GUIDE ◆

Location: Dinosaur National Monument, Northwestern Colorado

Peak Energy Time: Late afternoon sun, after thunderstorms, and near solstice sunsets

Energy Type: Fossil-laden sandstone, uplifted geologic time layers, uranium traces, and river-carved energetic flow

Best Activity: Fossil meditation, long-form journaling, and deep-time perspective practice

Symbol: Time Vault – Represents the preserved pulse of ancient life and access to Earth's memory grid

This landscape is a **living archive**, composed of uplifted and tilted sandstone, mudstone, and shale, many of which date back over 150 million years. The bones of long-extinct creatures rest in situ, embedded like sacred runes in the Earth, emitting a kind of **primordial frequency**—the residue of ancient biological energy fossilized into stone. Some intuitive visitors report sensations of awe, vertigo, or deep memory while standing in the quarry walls or hiking along the Morrison Formation exposures. It is not simply the knowledge of the past, but the presence of it, that lingers here.

Geologically, Dinosaur National Monument sits within a zone of **stratigraphic uplift** and erosion that reveals layer upon layer of sedimentary history. But beyond its fossil fame, the region also holds **uranium-bearing deposits**, adding to the subtle radiative pulse and magnetic complexity of the land. These mineral signatures—along with the high silica content of the sandstone and the energetic movement of the river corridors—create a kind of **geoelectric symphony**: low-frequency, steady, and ancient.

The **confluence of rivers** plays a crucial role in this hotspot's energy profile. As water cuts through stone, it not only reshapes land—it energizes it. Flowing water over mineral-rich beds generates subtle electric currents and negative ions, which may help explain the invigorating, almost electric feeling that some experience when hiking near the Echo Park or Whirlpool Canyon areas.

Dinosaur National Monument is not loud in its energy. It's deep, quiet, and vast—like time itself. Those sensitive to Earth's vibrational fields may find themselves slipping into altered states of awareness, especially when sitting among the stone layers at dusk or wandering the desert washes at dawn. This is a place of **perspective**, of **integration**, where the ephemeral nature of human life meets the enduring, grounded hum of deep time.

Whether you come seeking prehistoric wonder, geologic clarity, or a moment to connect with the planet's oldest memories, Dinosaur National Monument offers an invitation to **slow down, sink in, and listen** to the Earth's ancestral voice echoing through stone.

ICE LAKE BASIN

CRYSTALLINE THRESHOLD IN THE SKY

High above the town of Ouray, tucked deep within the San Juan Mountains, lies a hidden amphitheater of light and water—Ice Lake Basin. At over 12,000 feet in elevation, this otherworldly alpine cirque is a kaleidoscope of wildflower meadows, granite spires, and fluorescent turquoise lakes fed by glacial melt. But beyond its postcard perfection, Ice Lake Basin hums with an energetic resonance that is both ancient and luminous, like the Earth exhaling in clarity.

◆ ENERGY SIGNATURE GUIDE ◆

Location: Ice Lake Basin, Ouray, CO

Peak Energy Time: Sunrise, After Thunderstorms, and Late Summer Bloom

Energy Type: Glacial-fed turquoise lakes, volcanic mineralization, quartz and feldspar-rich granite, high-altitude ionization

Best Activity: Crown chakra meditation, breathwork for elevation clarity, and stillness in the presence of water

Symbol: ◇ Sky Mirror – Represents crystalline thought, threshold awareness, and light-coded renewal

This basin is not simply beautiful—it's **electrifying** in its serenity. The moment you crest the final switchbacks and step into the upper basin, there is a perceptible shift in atmosphere. The thin air amplifies subtle perception. The sounds fall away. The colors intensify. What remains is a shimmering stillness, a kind of vibrational field shaped by sky, water, and crystalline stone.

Geologically, Ice Lake Basin sits atop a heavily mineralized section of the San Juan Volcanic Field, a region shaped by immense explosive eruptions during the mid-Tertiary period. The rugged walls surrounding the basin are remnants of ancient calderas—collapsed volcanic chambers now carved into sweeping cliffs, spires, and ridgelines. These formations are rich in silica, feldspar, quartz, and trace metals like silver and copper. Some zones even contain minor occurrences of pyrite and magnetite, adding to the region's **subtle conductivity** and **magnetic complexity**.

The lakes themselves, with their surreal blue hues, are colored by **rock flour**—fine glacial silt suspended in meltwater. These suspended particles refract light in the higher spectrum, creating a luminous field of **turquoise and cobalt** that seems to radiate energy. Water, as always, is not just a physical presence but an **energetic conduit**—a moving mirror that reflects, amplifies, and subtly tunes the electromagnetic field of the surrounding space.

At this elevation, **ionization from atmospheric pressure gradients** and rapidly changing weather can also affect perception and physiology. Storms build quickly in the San Juans, and the electric charge in the air before a lightning strike is unmistakable. Hikers often report a buzzing sensation on their skin, or a strange inner clarity—like being washed clean by the sky. This is not a coincidence. **Electromagnetic charge build-up in high alpine zones** interacts with the body's bioelectric system, and in sensitive individuals, can trigger heightened awareness, subtle anxiety, or even visionary states.

But Ice Lake Basin isn't just high—it's **highly attuned**. The entire region acts like a satellite dish for energy: shaped by geological uplift, energized by volcanic mineralization, and acoustically reflective thanks to the stone amphitheater configuration. Sounds carry strangely here—sometimes seeming to echo inward, then dissolve into pure silence. This **acoustic stillness** creates ideal conditions for **resonant breathwork, mantra, or internal dialogue**.

From an energetic perspective, Ice Lake Basin feels like a **threshold**—a liminal space between dimensions, where the upper chakras naturally open. Crown and third eye energy centers may activate spontaneously here, not through effort, but through attunement. The sheer altitude clears the lower noise. What remains is subtle, crystalline, elevated.

For many Indigenous cultures and mystics across traditions, **high alpine basins** have long been considered **spirit sanctuaries**—places for pilgrimage, prayer, and communion. Ice Lake Basin fits this archetype perfectly. It's a site of both internal reflection and external awe, where the veil between human time and geological time thins. And while the views are dramatic, the energy is anything but flashy. It's quiet. Clear. Unwavering.

Whether you approach it as a physical challenge or a sacred journey, Ice Lake Basin rewards stillness more than striving. The path upward can be grueling—nearly 8 miles round-trip with over 2,500 feet of elevation gain. But the energy at the top **greets you like a ceremony**. A baptism in sky-water. A download from the stone itself.

This is a place to **listen to the wind**, to let go of questions and receive clarity in return. Bring your breath. Bring your silence. And let the basin show you what stillness at 12,000 feet can unlock within.

WHISPERING SPIRES OF ASH AND TIME

Tucked deep within the La Garita Wilderness of the San Juan Mountains, far from paved roads and cell signals, lies a place that feels like another planet—or perhaps a forgotten dream of this one. Wheeler Geologic Area is remote, silent, and utterly surreal. A crown of eroded volcanic spires, pinnacles, and hoodoos rise like sentinels from the earth, carved by wind and time from soft ash and tuff. It's not just geology—it's a sculpture garden of memory, formed from fire, worn by water, and whispering with quiet, ancient power.

◆ ENERGY SIGNATURE GUIDE ◆

- **Location:** Wheeler Geologic Area, San Juan Mountains, CO
- **Peak Energy Time:** Midday sun, after summer storms, and early fall light shifts
- **Energy Type:** Ash-flow tuff spires, eroded volcanic hoodoos, silica-charged formations, and subtle geomorphic memory
- **Best Activity:** Visual meditation, shadow work, and stone-figure contemplation
- **Symbol:** Silent Sentinel – Represents ancient transformation, elemental memory, and stillness through erosion

Getting to Wheeler isn't easy. The standard access route is a 14-mile round-trip hike, or a 4WD-only road that twists and climbs through the San Juans like a pilgrimage path. This effort, however, is part of the place's energetic charm. It requires intention. It slows the mind. And by the time you arrive, the stillness has already begun working on you.

Wheeler is the remnant of one of the largest volcanic eruptions in Earth's history—**the La Garita supervolcano**, which erupted over 27 million years ago. The landscape you walk through is composed almost entirely of **ash-flow tuff**, a dense rock formed from compressed volcanic ash that blanketed the region in the wake of the eruption. Over millennia, wind, rain, and freeze-thaw cycles sculpted this ash into strange, elegant forms—**pillars, fins, arches, and narrow towers** that seem to hum with the residue of elemental upheaval.

There is a sacred geometry in their arrangement. Some formations resemble seated figures, others look like rising serpents, fractured altars, or wind-carved cloaks. Hikers have reported seeing faces, eyes, hands reaching skyward—though these visions shift and change with the light. The **morphic language** of Wheeler is subtle but unmistakable. It's a place that teaches you how to see again.

Energetically, Wheeler pulses with **stilled transformation**. Unlike more active volcanic zones filled with geothermal heat and electromagnetic buzz, Wheeler is **cool, contemplative, and carved down to essence**. The energy here is not about ignition—it's about what remains after fire has passed through. It's about shape-shifting. About remembering who you were after the story burns away.

From a geophysical perspective, Wheeler holds **high silica content** from its volcanic origin, which contributes to its pale color and dry acoustic quality. Sound behaves differently here—**absorbed by the soft stone, diffused by the labyrinthine walls**. This creates an uncanny silence, a sonic compression chamber that heightens awareness. It's an ideal place for **quiet visualization, walking meditation, or even shadow work**, as the eerie stillness makes the inner voice easier to hear.

There are no hot springs here, no lush forests, no electrical hums—just the **wind, the stone, and the shape of deep time**. That simplicity is what makes it so powerful. The spires act like **amplifiers for contemplation**, each one a frozen exhale of the Earth's past breath. You are walking through the ghost of fire, and the ghost is still speaking.

And yet, for all its alien beauty, Wheeler is grounded. It teaches humility, presence, and trust. The hike in demands pacing. The weather demands respect. The formations demand stillness. Many who visit describe a sensation of **being watched**, but not in an ominous sense—more like being welcomed into a gathering of ancient teachers. There's intelligence in the stone here, formed not of logic but of pattern, shape, and silence.

There's also a kind of **inner cleansing** that occurs at Wheeler. The ash rock itself is fragile, light, easily worn away—and in its presence, the heaviness of thought, ego, and expectation often erodes too. You find yourself standing taller. Breathing more deeply. Thinking less. The mind quiets as the body realigns with the rawness of the place.

Wheeler Geologic Area is not just a geological curiosity. It's an **energetic sculpture**, an altar of transformation, and a cathedral of elemental erosion. It asks for nothing and gives everything, if you can stay long enough to receive it.

This is a place to listen—not with your ears, but with your whole field. To trace the stories the Earth once told in fire, now softened by air and time. To walk slowly, speak gently, and let the stone shape you in return.

CHICAGO LAKES TRAIL
GATEWAY OF ASCENT AND REFLECTION

Tucked into the high country west of Idaho Springs, the Chicago Lakes Trail begins like a whisper and ends in pure alpine exhale. This trail isn't just a path through trees and toward lakes—it's a gradual energetic unfolding, one that mirrors the internal process of elevation, transition, and integration.

◆ **ENERGY SIGNATURE GUIDE** ◆

Location: Chicago Lakes Trail, Mount Evans Wilderness, near Idaho Springs, CO

Peak Energy Time: Early morning fog, after summer monsoons, and near autumn equinox

Energy Type: Glacial-carved cirques, reflective alpine lakes, granite bedrock with quartz veins, and cascading water-channel energy

Best Activity: Flow-state hiking, grounding breathwork by the lower lake, and sky-gazing meditation at the upper lake

Symbol: Mist Gate – Represents transition, ascent, and emotional clarity through 73 movement

From the outset, the trail descends through pine and fir, hugging the steep canyon carved by Chicago Creek. The energy here is enclosed, quiet, almost womb-like. The sound of rushing water threads through the trees like a nervous system, and the trail—lined with granite boulders, moss, and shadow—asks you to move slowly, to listen carefully. It's not an overpowering energy, but a **deep, rooted pulse** that grounds you before your climb.

As the trail progresses upward and begins its steady switchbacks, you move through several **distinct energetic zones**. The first lake you encounter—the lower Chicago Lake—rests in a wide glacial bowl. It's ringed with granite cliffs and quartz-veined bedrock, where meltwater from the upper basin trickles down in shimmering cascades. This space has a **clean, open charge**. The kind that clears mental clutter and invites stillness. Here, the reflection of the sky in the water feels like a mirror not only of the land, but of the self. It's an ideal place for **grounding meditations, pranayama, or quiet journaling**, especially near sunrise or misty mornings.

Geologically, the Chicago Lakes Basin is part of the ancient **Mount Evans Batholith**—a mass of Precambrian granite formed more than a billion years ago. The granite here contains **abundant quartz**, which is both piezoelectric and symbolically linked to clarity, amplification, and spiritual resonance. As water flows over these rocks—charged by snowmelt and mineral runoff—it may subtly generate electrical microcurrents, contributing to the region's **natural geomagnetic hum**.

Ascending toward the upper lake, the environment opens dramatically. The trees thin, the air sharpens, and the sky seems impossibly close. Here, the trail shifts from an earthy, forested passage to a high alpine amphitheater. The wind picks up. The light intensifies. The silence deepens. This final leg of the trail is where many report feeling **energetic release, sudden insight, or emotional clarity**—not unlike the spiritual sensation of crossing a threshold.

Energetically, this trail is a **ritual of transition**. It begins in the low, closed space of the forest, dips into a reflective bowl, and culminates in a high-elevation basin perched beneath towering cliffs. In this way, the landscape echoes a classic symbolic journey: descent, reflection, and ascent. It's a physical **initiation pathway**, whether you frame it geologically or spiritually.

The lakes themselves are shaped by **glacial cirques**, naturally occurring energy concentrators formed by the slow carving of ice. These curved basins often act as **natural resonant chambers**, where sound and energy reflect inward. In metaphysical terms, they're places where the past has been worn away, revealing the essence beneath. Standing here, surrounded by alpine silence and ringed stone, one may feel time pause—or expand.

Beyond the geology and the altitude, there's something quietly sacred about Chicago Lakes. It doesn't have the grandeur of a national park or the raw magnetism of a volcanic zone, but it **gently aligns** the body and mind. It offers clarity not through intensity, but through structure: a perfect sequence of grounding, reflection, and release.

The final overlook—where the trail crests above the upper lake—is a place of **vantage and surrender**. You've climbed over 2,500 feet, passed through elemental layers, and now stand face-to-face with the sky. The energy here is light but penetrating, like sunlight filtered through ice. Some find inspiration. Others feel tears well up unexpectedly. Many simply sit in silence.

If you listen closely, you may notice the quiet electric hum of water over stone, or the wind echoing through quartz-laced cliffs. You might feel your heart slow to match the rhythm of the lake, or your thoughts pause long enough for something new to arrive. This is what makes the Chicago Lakes Trail special. It doesn't overwhelm. It **unfolds you**.

Whether you hike here seeking solitude, clarity, or communion with the alpine world, this trail offers an energetic mirror to the inner path. Come with your breath. Come with your stillness. And let the land show you how reflection, movement, and elevation are all parts of the same sacred climb.

LIMBER GROVE
THE ELDERS OF THE WIND

Perched on the flanks of Colorado's Mosquito Range, just outside of Fairplay, the Limber Grove Trailhead leads to a windswept sanctuary where time is written in twisted trunks and sun-bleached bark. Here, ancient limber pines and bristlecone pines—some more than 1,000 years old—stand as living monuments to endurance, memory, and transformation. This is not a loud place. It does not hum or surge like geothermal basins or magnetic rift zones. The energy here is older, quieter, and deeply grounding. It moves like breath through bone.

◆ ENERGY SIGNATURE GUIDE ◆

📍 **Location:** Limber Grove Trailhead, Mosquito Range, near Fairplay, CO

⏳ **Peak Energy Time:** Early morning stillness, before summer storms, and during the fall leaf drop

⚡ **Energy Type:** Ancient bristlecone and limber pine trees, weather-sculpted granite outcrops, high-elevation oxygen clarity, and subtle geomorphic wind harmonics

🔱 **Best Activity:** Slow tree-walk meditation, elder connection rituals, and breath-holding visualization near twisted trunks

🎯 **Symbol:** 🌲 Timewalker Spiral – Represents endurance, slow wisdom, and living memory through rooted resilience

The trail begins unassumingly—gentle incline, dusty path, a few scattered wildflowers clinging to the hillside. But soon you begin to notice the trees. At first, they seem oddly shaped. Twisted. Hunched. Then, with time, you start to recognize them as **individual presences**, each one a sculpture formed by **centuries of wind, lightning, drought, and high-elevation sun**. Their shape is not random—it's the result of a conversation with the elements. And if you walk slowly enough, that conversation begins to include you.

Geologically, the Limber Grove sits on a shoulder of **weathered granite bedrock**, rich in feldspar and quartz, overlain by thin alpine soils. The combination of **exposed rock, high elevation (over 11,000 feet), and consistent wind patterns** makes this region both ecologically extreme and energetically unique. There's a clarity here—not just in the air, but in the vibrational field. The trees, like antennae, seem to **channel and radiate** this stillness outward.

In subtle energetic traditions, **trees are recognized as timekeepers and connectors between worlds**—their roots deep in the Earth, their branches reaching into the sky. Nowhere is this more palpable than at Limber Grove. These trees are slow lightning. They've endured solar flares, temperature extremes, and geological shifts, storing in their spiral grain the **frequency of survival and adaptation**. To sit with one—really sit—is to place your body against a library of solar storms and star cycles.

Many visitors report feeling a **slowing of the nervous system**, a shift in breath rate, or an unexpected release of emotion when walking among the older trees. It's not unusual to feel watched here—not by animals or spirits, but by the landscape itself. The grove holds presence. It has awareness. It has seen everything, and it asks for reverence in return.

This is not a power spot in the usual sense. There are no electromagnetic anomalies or hot springs, no towering peaks or carved canyons. But Limber Grove is a **portal of subtle memory**. The pines here live slowly, and they invite you to do the same. Their bark is worn smooth by wind, their trunks curled by unseen forces. And in the presence of this **elemental sculpting**, your own patterns—mental, emotional, energetic—begin to show themselves. You start to feel where you resist the wind. Where you stiffen. Where you refuse to adapt.

Energetically, this area aligns with the **root and heart chakras**. The grounding comes from the bedrock and rootedness of the trees. The heart-opening emerges through stillness, breath, and the way light falls on the bark in late afternoon. **Fall is especially powerful here**, when golden leaves drop like prayers and the land quiets itself before the snow.

There is also a kind of **bioelectric resonance** at play. Limber and bristlecone pines contain resin canals and water-storing structures that allow them to survive extreme drought. This internal architecture, along with their twisting grain and persistent exposure to wind, may create low-level piezoelectric effects. Combined with the **quartz-flecked granite beneath**, the area may subtly **enhance grounding, emotional release, and energy clarity**—especially for those sensitive to subtle fields.

This is a place for **walking slowly**, for listening to what bends and what breaks. For honoring the long view. It's an ideal site for **breathwork, grief rituals, elder connection meditations**, and simply being with your own stillness. Nothing here moves quickly—but everything here moves with purpose.

Whether you arrive at Limber Grove looking for ancient trees or unexpected insight, the land meets you with quiet precision. You may not even realize the shift until you return to the world below—and find your own rhythm has changed.

You have moved through the **elders of the wind**, and they have shaped you.

WEST MAROON PASS
THE BLOOMING THRESHOLD

Rising from the wildflower-laced meadows outside of Crested Butte and climbing toward the ragged divide of the Elk Mountains, the West Maroon Pass Trail is less a hike and more a rite of passage. Traversing this trail is to move through a seasonal portal, where geology, bloom, and elevation converge to create an energetic field that is alive, expansive, and deeply transformative.

◆ ENERGY SIGNATURE GUIDE ◆

Location: West Maroon Pass Trail, between Crested Butte and Aspen, CO

Peak Energy Time: Sunrise in peak wildflower season, after rain, and during autumn gold light

Energy Type: Volcanic and sedimentary layering, high alpine wildflower charge, quartz-rich scree fields, and rhythmic elevation flow

Best Activity: Flow-state hiking, floral breathwork, heart-opening rituals at the pass

Symbol: 🌼 Blooming Threshold – Represents seasonal vitality, emotional expansion, and the natural arc of ascent and release

The route itself begins humbly—meadows dotted with lupine and paintbrush, pine-scented breezes, and creeks weaving quietly beneath wooden bridges. But as the miles unfold and the grade begins to climb, the landscape opens. What begins as a grounded valley trail gradually becomes a **processional path through a living cathedral of flowers and stone**.

In mid-to-late summer, this section of the Elk Mountains explodes with wildflowers. Not just a few, but thousands—dense tapestries of color carpeting the alpine floor. Columbine, bluebells, monkshood, gentian, and fireweed bloom in waves, painting the landscape in a vibrational frequency that can feel almost euphoric. This is more than beauty. It's **biodynamic energy**—floral charge activated by sunlight, water, and soil chemistry.

Flowers are not inert. They're subtle energetic transmitters. And in West Maroon Basin, where mineral-rich volcanic and sedimentary layers intermingle beneath the surface, this transmission becomes palpable. Many hikers report sensations of joy, creative inspiration, or sudden emotional release while moving through the floral corridor. This is a place where the **heart field naturally opens**—without ceremony, without effort.

Geologically, the region is a **collision zone of time and transformation**. The peaks and ridgelines around West Maroon are composed of both **sedimentary layers from ancient seabeds** and **volcanic tuffs and breccias** left behind by explosive eruptions. This interbedding creates subtle conductivity contrasts beneath your feet—zones of differing mineral charge and grounding potential. The trail winds between these layers like a thread through a tapestry, allowing you to **move across energetic boundaries without even realizing it**.

By the time you reach the final approach to West Maroon Pass, the trail has done its work. You've climbed through shifting terrain, crossed snowfields and waterfalls, and inhaled the breath of the mountain itself. Your breath deepens. Your thoughts quiet. And then—you arrive.

The **pass** itself is a narrow saddle between two great basins, a wind channel carved by glaciers and time. The view opens in both directions—toward Aspen on one side, and Crested Butte on the other. It feels like standing at the **meeting point of worlds**. In fact, it is.

This ridgeline is a **transitional energy point**—a classic "third chakra to fourth chakra" movement. The effort to reach it is physical, solar, and will-based (solar plexus). But the view from the top, and the field it holds, is purely heart-centered. It's where effort turns into awe. Where tension releases into breath. Where clarity becomes more than an idea.

Energetically, this area is also influenced by **high-altitude ionization** and **wind pattern compression**. As air rushes through the saddle, it stirs not just dust and scent, but subtle shifts in pressure and electromagnetic charge. Sensitive individuals may feel this as tingling skin, light-headedness, or an expanded awareness that seems to echo through the entire basin. It is the body adapting to altitude—but also attuning to frequency.

Some speak of visions at the pass. Others feel guided messages or ancestral presence in the wind. These reports are not unusual in thin-air places shaped by **water, mineral, and seasonal motion**. West Maroon doesn't shout—it opens. It doesn't demand—it invites.

For those who hike the full route from Crested Butte to Aspen, the journey becomes a symbolic **traverse of transformation**. You begin in bloom, pass through exposure, and arrive in revelation. And the trail itself—gentle at times, demanding at others—mirrors the emotional rhythm of any inner ascent.

Whether you're here for wildflowers, high-alpine beauty, or the sheer joy of movement, West Maroon Pass offers more than elevation. It offers **alignment**. It offers a walkable metaphor for change, for flow, for the rewards of patient ascent.

This is a trail that **blooms to life** in just a few short weeks each year. But what it opens in the human field—**joy, clarity, release, breath**—can linger long after the petals fade.

Come for the flowers. Stay for the light. Leave transformed.

GATEWAY INTO THE UPPER REALMS

Tucked into the wild alpine folds of the Indian Peaks Wilderness, just outside the quirky mountain town of Nederland, the 4th of July Trail is more than a seasonal hiking destination. It is a threshold trail—a living passageway through stone, water, and wildflowers that seems to mirror the process of shedding what is no longer needed and stepping into clarity. While many hikers rush toward its high alpine lakes, those attuned to energy will feel the invitation much earlier: to slow down, to listen, and to move in rhythm with the unfolding terrain.

◆ ENERGY SIGNATURE GUIDE ◆

- **Location:** 4th of July Trail, near Nederland, Colorado
- **Peak Energy Time:** Early morning during snowmelt, midsummer bloom, and evening alpenglow
- **Energy Type:** Glacial valley flow, granite and gneiss bedrock, waterfall-charged air, and wildflower-enhanced field resonance
- **Best Activity:** Morning movement meditation, breath-alignment near waterfalls, and intuitive journaling above treeline
- **Symbol:** Alpine Portal – Represents seasonal emergence, inner vitality, and the crossing into clarity

Beginning at the **4th of July Trailhead**, this path immediately introduces a dual sensation: **upward movement and subtle containment**. The narrow trail rises through dense conifers, rocky cutbacks, and moss-laced granite boulders. This is an **energetic compression zone**, where the senses sharpen and the mind begins to quiet. Surrounded by granite and gneiss—ancient, crystalline, pressure-forged bedrock—you feel the magnetic density beneath your feet. This is Earth's root chakra terrain, and it creates a natural container for inward turning.

Soon, the forest begins to loosen. The sound of water becomes constant—runoff from snowfields melting into tumbling creeks and waterfalls. Water **weaves movement through the stone**, threading vitality through the landscape. These streams are more than features; they're **energetic guides**. Negative ions in the rushing water may uplift the mood, soothe the nervous system, and open the breath. A few moments near these cascades, especially in early morning, can shift your entire physiological state.

As you ascend, the terrain softens again into **open alpine meadows**, especially magical in July and August. This is when the trail lives up to its name: it blooms like a celebration. Wildflowers burst in waves across the valley floor—Indian paintbrush, bluebells, asters, columbines, and elephant heads dance in light, adding **color frequency to the energy field**. These natural bloom fields aren't just visually beautiful—they generate vibrational uplift. Many hikers report a sense of lightness, elation, or heart-centered clarity while walking among the flowers. This part of the trail corresponds energetically to the **heart chakra**—expansive, full of breath, and alive with resonance.

The 4th of July Trail leads to multiple destinations—Diamond Lake, Arapaho Pass, and beyond—but all of them pass through the same energetic arc: **forest containment, water movement, floral expansion, and alpine clarity**. The transition from one zone to another feels ritualistic, like moving through a **series of inner thresholds**. Each step into higher elevation corresponds to a subtle shift in awareness. This is why many people leave this trail feeling clearer, calmer, or even subtly altered—even if they wouldn't use those words themselves.

Geologically, the Indian Peaks region is a **glacially carved sanctuary**—a remnant of Ice Age forces that sculpted deep valleys and jagged peaks. The **granite and gneiss underfoot are some of the oldest rocks in Colorado,** forged under immense heat and pressure in the Earth's deep crust. These rocks, particularly those containing feldspar and quartz, may exhibit subtle piezoelectric properties—meaning they can generate tiny electric charges under mechanical stress. While this effect is invisible to the eye, its presence may contribute to the **tingling, grounding sensation** some feel when walking across this terrain.

Energetically, the 4th of July Trail resonates with movement and flow. Unlike volcanic landscapes that hold intensity or canyons that echo ancient stillness, this trail encourages you to move forward while staying connected. It is ideal for **walking meditation, breath practices, or mantra repetition**—especially in the spaces between tree and sky, where wind passes through like a guide.

Toward the upper reaches of the trail—whether you choose to summit a pass or simply pause at a lake—you'll notice a shift. The wind carries differently. The trees are gone. The sky presses closer. This is the **alpine crown** of the trail, where thoughts dissolve and presence sharpens. Many find this space ideal for **intuitive journaling, open-ended contemplation, or sitting in raw stillness**. It's not about answers here—it's about alignment.

As you descend, the trail offers one final gift: integration. What you passed through on the way up—forest, stream, bloom—becomes a mirror. You carry the energy now. The clarity isn't just in the landscape—it's in you.

SNIKTAU TO GRIZZLY
THE RIDGE OF INSIGHT AND INTEGRATION

Straddling the Continental Divide just west of Loveland Pass, the ridgeline route from Mount Sniktau to Cupid Peak to Grizzly Peak D is more than an alpine traverse—it's a wind-shaped corridor of clarity, recalibration, and elemental initiation. Often underestimated due to its accessibility, this high-elevation walk is a masterclass in energy layering, spatial awareness, and the internal stillness that emerges when the rest of the world falls away.

◆ ENERGY SIGNATURE GUIDE ◆

Location: Mount Sniktau, Loveland Pass, CO

Peak Energy Time: Sunrise, midwinter inversions, and solar flares

Energy Type: Exposed ridge energy, fast wind alignment, high-elevation ionization, and quartz-flecked granite base

Best Activity: Wind-breath meditation, vision clarity journaling, and third eye alignment at summit

Symbol: 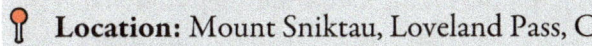 Sky Spine – Represents intuition, insight, and elemental purification

Beginning from the parking lot at over 11,900 feet, the Sniktau trail wastes no time. Within minutes, you're above the trees, surrounded by endless sky, battered by wind, and walking on stone. This sudden exposure is not just visual—it's vibrational. The entire environment becomes a tuning fork. **Your thoughts sharpen. Your breath changes. Your awareness expands.**

Mount Sniktau: Crown and Clarity

Mount Sniktau rises quickly, a sharp, curving ridge leading to a summit just above 13,200 feet. It's known for its commanding views and proximity to Denver, but beneath the beauty is a powerful energetic signature: **clean, elevated, and piercingly clear**. The wind here is constant, stripping away distraction and unnecessary thought. The quartz and feldspar in the underlying granitic rock contribute to a kind of piezoelectric clarity—a low-frequency pulse that aligns easily with the third eye and crown chakras.

Many visitors describe Sniktau as a place of **vision**—not just in terms of landscape, but of internal understanding. It's a site well-suited for **intuitive journaling, breathwork, or decision-making rituals**, especially at sunrise, when the land glows pink and gold beneath the rising light. Here, elevation is insight.

But Sniktau doesn't invite you to linger long. The wind keeps moving—and so do you.

Cupid Peak: The Saddle of Balance

◆ **ENERGY SIGNATURE GUIDE** ◆

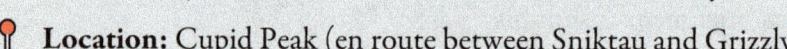

📍 **Location:** Cupid Peak (en route between Sniktau and Grizzly)

⏳ **Peak Energy Time:** Late morning stillness, spring melt, and during atmospheric shifts

⚡ **Energy Type:** Gentle saddle convergence, flow between peaks, subtle energy cycling between earth and sky

⚗ **Best Activity:** Heart-breath connection, movement-based prayer, and horizon-focused intention setting

🎯 **Symbol:** 💗 Open Pass – Represents energetic balance, openness, and rhythmic alignment

The trail descends from Sniktau and threads through a gentle saddle before rising again toward **Cupid Peak**. Though modest in elevation gain, this transition holds an important energetic shift. The saddle acts as a **valley of breath**,

a moment of recalibration between the sharper frequencies of Sniktau and the broader resonance of Grizzly.

Cupid Peak, despite its soft name, offers a **heart-centered anchoring**. The exposed ridgeline invites expansive breathing, and the energy here is rhythmic—**a pulsing equilibrium** between effort and ease, insight and embodiment. Hikers often find themselves walking at a steadier, more natural pace along this stretch, feeling grounded and open.

This is a perfect place for **movement meditation**, for synchronizing footsteps with breath, or for setting intentions carried on the wind. The visibility in all directions allows for a sense of connection—not just to place, but to the whole of your path. In this way, Cupid functions like an energetic fulcrum, bringing you back to the body between the mind of Sniktau and the depth of Grizzly.

Grizzly Peak: The Quiet of Completion

◆ **ENERGY SIGNATURE GUIDE** ◆

Location: Grizzly Peak D (13,427 ft),   Summit County

Peak Energy Time: Solar noon, after lightning activity, and during late-season solitude

Energy Type: High-alpine magnetism, glacial memory in cirque walls, deep rock echo and silence compression

Best Activity: Summit stillness, grounding breath-holds, and integration meditations

Symbol: Ancient Ridge – Represents clarity through endurance, wideview wisdom, and breath-carved strength

Grizzly Peak D, often overlooked in favor of flashier 14ers, is a **summit of integration**. At 13,427 feet, it feels massive—its broad shoulders and quiet presence standing in contrast to the sharpness of Sniktau or the softness of

Cupid. The final approach is steep, often snow-covered well into summer, and demands presence.

Here, the wind changes again—lower, quieter, more deliberate. The energy of Grizzly Peak is **deeply grounding**, like a granite root reaching into the mountain's memory. It is a place of **completion, of reflection, and of breath-based stillness**. If Sniktau opens the mind and Cupid balances the heart, then Grizzly **anchors the journey into the bones**.

The surrounding cirques and shadowed valleys hold glacial remnants—ghosts of ice that shaped the landscape and left behind **natural memory chambers** in the stone. Sitting here, you may feel time stretch. Or collapse. Or melt entirely into the moment. This is not a peak that demands movement. It invites you to **sit, listen, and feel what remains after all the effort.**

🐾 A Ridgewalk of Realignment

What makes this route so powerful is not the altitude alone, but the **sequence of energies**. You begin sharp, elevated, and wind-cleansed on Sniktau. You pass through openness and expansion on Cupid. And you arrive at depth, rootedness, and raw presence on Grizzly.

This is more than a hike—it is a **moving meditation through elemental chambers**. Each step along the ridgeline offers both a challenge and a clearing. And when you return, whether by backtracking or looping, you carry not just the physical accomplishment, but an **energetic reset**: mind sharpened, heart balanced, body grounded.

Whether you're walking through weather, through yourself, or simply through space, the Sniktau–Cupid–Grizzly route offers a rare kind of gift—a **conversation with wind, stone, and sky** that leaves you lighter, clearer, and more wholly aligned.

HANGING LAKE

SUSPENDED STILLNESS AND THE STONE THAT BREATHES

Tucked into a narrow cleft high above Glenwood Canyon, Hanging Lake feels like a secret held aloft by time. Crystal-clear and impossibly still, this turquoise pool seems to float between earth and sky—fed by waterfalls, cradled by cliffs, and framed by moss, travertine, and stone. It is one of Colorado's most iconic destinations, but beneath its Instagram-famous beauty lies a deep geological and energetic intelligence—a place formed by tectonics and uplift, mineral reaction, and slow, sacred flow.

◆ ENERGY SIGNATURE GUIDE ◆

- **Location:** Hanging Lake, Glenwood Canyon, CO
- **Peak Energy Time:** Mid-morning light, post-rainfall, and during spring melt
- **Energy Type:** Travertine terraces, suspended turquoise pool, falling water ionization, limestone bedrock, and chlorophyll resonance
- **Best Activity:** Water-focused meditation, emotional cleansing breathwork, and heart-focused journaling
- **Symbol:** 💧Suspended Mirror – Represents elevation through surrender, inner clarity, and purification by beauty

The trail to Hanging Lake begins at the canyon floor, where the roar of the Colorado River reverberates through ancient limestone walls. As you ascend the rocky path, the environment shifts—from lowland riparian to high cliff garden in under two miles. This **vertical journey mirrors the energetic arc** of the lake itself: grounded in movement, uplifted in stillness.

When you arrive, Hanging Lake reveals itself suddenly—an aquamarine basin clinging to a shelf carved into the limestone cliffs. Water flows in delicate strands from the spongy moss above, cascading from **Spouting Rock and Bridal Veil Falls**. Below the surface, the water is so clear that logs resting on the lakebed appear suspended in air. Everything here suggests **weightlessness**, even though its creation is rooted in stone.

The Geology of Hanging Lake

Hanging Lake is a rare example of a **travertine lake**, formed not by glacial carving or tectonic faulting alone, but through a **geochemical process involving water, rock, and time**. The story begins with the **Leadville Limestone**, a dense marine deposit laid down more than 300 million years ago. This limestone is rich in calcium carbonate, which plays a key role in the lake's formation.

As groundwater flows through fractures in the surrounding rock, it dissolves calcium carbonate from the limestone. When that mineral-rich water reaches the surface and is exposed to air, the carbon dioxide in the water escapes, causing the calcium to **precipitate and crystallize**. Over thousands of years, this slow chemical reaction forms **travertine**—a porous, honey-colored stone that grows in terraces, mats, and rims around the pool.

Hanging Lake rests on a natural travertine bench that continues to build and expand with time. The lake itself is essentially a **living stone formation**, shaped by slow-moving water chemistry and biological activity, including algae and moss that trap and hold mineral particles. The result is a **place that is both stone and water, motion and stillness, geology and life**.

This geologic cradle is further supported by the **uplifted sedimentary layers** of Glenwood Canyon—one of Colorado's most striking geological formations. The canyon is composed of stacked strata, including sandstone, limestone, and shale, all tilted and lifted by the **Laramide Orogeny**, a period of mountain-building that raised the Rockies roughly 70 million years ago.

The trail to Hanging Lake passes through these ancient layers, crossing fault lines, seeps, and fracture zones—**each one subtly influencing water flow, mineral content, and energetic tone.**

Water and Resonance

Energetically, Hanging Lake is a place of **emotional suspension**. The travertine creates an acoustic buffer—absorbing sound and softening movement. The waterfalls generate a **fine mist of negative ions**, which have been shown to improve mood and nervous system function. Together, these elements form a microclimate of **gentle energetic cleansing**—ideal for emotional release, breathwork, or quiet contemplation.

The water itself seems to hold memory. The logs submerged in the lake are decades old, perfectly preserved due to the high mineral content and cold temperature. These silent forms rest in a kind of **liquid stillness**, echoing the invitation Hanging Lake offers to its visitors: **Pause. Reflect. Surrender to what is held, not pushed.**

Hanging Lake aligns energetically with the **heart and throat chakras**. The turquoise hue of the water resonates with the high-heart frequency—a bridge between expression and feeling. It is a place to speak softly, to feel fully, to let clarity rise from quiet.

Ritual in Stillness

Many visitors find themselves whispering upon arrival, as though entering a natural cathedral. The stone amphitheater that surrounds the lake reflects sound and presence inward, creating a sense of **containment and echo**—not only of sound, but of intention. Sitting by the lake in stillness, especially near sunrise or misty mornings, can produce a state of gentle altered awareness: not vision, but **heightened perception**. A remembering of water, breath, and quiet power.

The return trail carries this clarity back down the mountain. What felt like a climb on the way up becomes a slow descent into integration. The body is lighter. The breath is deeper. Hanging Lake, though small in size, leaves a large imprint on the internal field.

This is not just a destination—it is a **lesson in suspension, chemistry, and surrender**. A place where rock becomes water, and water becomes mirror.

SPIRITUAL PRACTICES AND ENERGY

TO SUPERCHARGE YOUR SOUL AND SYNC WITH NATURE

Harnessing Earth's energy for deeper focus, healing, and transformation

Each power spot offers a unique energetic frequency, amplifying your spiritual practices. From grounding poses to intentional breathwork, learn to harness the Earth's energy for unparalleled clarity and peace.

Practicing meditation and yoga in nature has long been recognized as a way to deepen one's connection with the universe. The natural energy of places like Colorado's power spots can elevate spiritual practices to new heights, creating an immersive experience that aligns body, mind, and spirit with the forces of the Earth. Here, we explore the most effective techniques to attune yourself to these energies, allowing you to harness their full potential.

Grounding Practices: Connecting with the Earth

Grounding, also known as earthing, is the practice of making direct contact with the Earth's surface to stabilize the body's electrical charge. Many power spots in Colorado, particularly those with high mineral content, amplify this effect. Walking barefoot on rock formations, soil, or grass in places like Garden of the Gods or Lookout Mountain can enhance your energetic stability, providing a profound sense of connection and renewal.

Techniques:

Standing Meditation: Stand with your feet firmly planted on the Earth, close your eyes, and visualize energy moving up from the ground into your body. Breathe deeply and allow yourself to feel rooted.

Tree Pose (Vrikshasana): This yoga pose, practiced in a power spot, strengthens balance while aligning your energy with that of the natural world.

Crystal Charging: Bringing grounding stones like hematite, black tourmaline, or red jasper to an energetically charged location can amplify their properties, helping you stay centered.

Breathwork: Enhancing Your Connection to Energy Fields, is an essential tool for syncing with the frequencies of power spots. The way we breathe influences our energy, allowing us to absorb and harmonize with the natural electrical activity of the land.

Techniques:

Pranayama (Controlled Breathing): Practicing techniques such as Nadi Shodhana (alternate nostril breathing) can balance your body's electromagnetic fields, especially in areas with high telluric currents.

Deep Diaphragmatic Breathing: Standing on a high-energy location, take slow, deep breaths, feeling the expansion in your lungs. This enhances oxygen flow and aligns your energy with the space.

Fire Breath (Kapalabhati): A rapid breathing technique that invigorates the mind and body while stimulating internal energy circulation.

Meditation Practices to Absorb Energy

Meditation in nature's power spots allows for a deeper connection with the Earth's energy. Some locations, such as the San Luis Valley and Mount Princeton Hot Springs, have natural resonances that enhance spiritual experiences.

Techniques:

Guided Visualization: Imagine golden light surrounding you, pulling energy from the Earth and sky, syncing with the land's electromagnetic frequencies.

Sound Meditation: Using singing bowls or chanting mantras near rock formations, waterfalls, or geothermal springs can harmonize energy flow.

Silent Stillness: Sitting quietly in an energetically charged location, focusing solely on breath and sensation, allows you to absorb the vibrational frequencies of the land.

Yoga Practices to Amplify Energy Flow

Practicing yoga in power spots enhances postures and deepens your alignment with the forces of nature.

Recommended Poses:

Mountain Pose (Tadasana): Performed at high-altitude locations like Lookout Mountain to align with the sky and Earth simultaneously.

Downward Dog (Adho Mukha Svanasana): Encourages blood flow and stability in places of shifting energy fields.

Sun Salutations (Surya Namaskar): Practicing a flowing sequence in places where sunlight strikes rock formations increases vitality and balance.

Best Practices for Spiritual Energy Work

1. **Visit During High-Energy Times:** Equinoxes, solstices, and lunar phases impact electromagnetic fields, making these optimal times for meditation and yoga.
2. **Use Natural Elements:** Incorporate water, crystals, and incense to enhance your practice.

3. **Set an Intention:** Before beginning, set a focused intention to align with the energy of the space.
4. **Listen to Your Body:** Some locations may be overwhelming at first. Stay attuned to your responses and take breaks as needed.

By integrating meditation, yoga, and breathwork into Colorado's power spots, you can unlock new levels of consciousness, insight, and healing. These landscapes are not just places of beauty but also portals to deeper awareness and transformation.

CELESTIAL RHYTHMS, PLANETARY MOVEMENTS, AND EARTH'S SECRETS

Using solstices, equinoxes, and lunar phases to amplify energy alignment

Timing is everything. The Earth's energy shifts with the time of day, season, and celestial events. Discover the profound significance of dawn, midnight, solstices, and equinoxes for your energetic practices.

The Daily Energy Cycles: Harnessing the Flow of Time

The Earth breathes in cycles, and so do we. Each time of day holds a unique energetic signature that influences spiritual practices and energy absorption. Understanding these fluctuations allows you to align yourself with natural forces and maximize the impact of your meditation, yoga, or energy work.

Dawn: The Birth of Light and New Beginnings

Dawn marks the transition from night to day, a time of awakening and renewal. As the sun rises, the Earth's magnetic field stabilizes, and the natural electric currents running through the planet begin to shift. This period is ideal for setting intentions, meditation, and breathwork to welcome new energy into your life.

Best practices during dawn:

Sun Salutations (Surya Namaskar): Practicing yoga as the first light touches the Earth connects you with solar energy.

Morning Meditation: As the world awakens, tuning into the quiet energy of dawn enhances clarity and intention-setting.

Grounding Rituals: Walking barefoot on the earth, particularly on dewy grass, can recharge your energy field.

Midday: The Peak of Power and Manifestation

As the sun reaches its zenith, energy surges across the landscape. Midday is a time of heightened vitality, making it an excellent period for productivity, movement-based meditation, and engaging in physical energy work.

Best practices during midday:

Dynamic Yoga Practices: Power yoga or vinyasa flows harness the peak solar energy.

Breathwork Techniques: Engaging in rapid breathing exercises like Kapalabhati (fire breath) can help distribute high energy.

Manifestation Practices: Writing down goals, visualizing success, or performing energy-raising rituals aligns with the solar peak.

Sunset: The Transition of Light and Shadow

Dusk brings a cooling down, both in temperature and energy. It is a time for reflection, gratitude, and releasing what no longer serves you. The fading light offers a moment of introspection, and the shifting electromagnetic balance makes it an ideal time for balancing practices.

Best practices during sunset:

Gentle Yoga Sequences: Restorative yoga or yin yoga helps ease the body into relaxation.

Journaling and Reflection: Writing about the day's experiences while the sky shifts colors enhances clarity and self-awareness.

Cleansing Rituals: Burning sage or palo santo can clear residual energy before the night.

Midnight: The Depths of Intuition and Subtle Energies

Midnight is the hour when energy turns inward. The absence of sunlight heightens sensitivity to subtle vibrations, making this time ideal for deep meditation, astral work, and exploring dream states.

Best practices during midnight:

Lucid Dreaming Practices: Keeping a dream journal helps track messages from the subconscious.

Stillness Meditation: Sitting in complete silence at this time allows deep internal exploration.

Energy Healing: Reiki and other forms of healing work are amplified during this time.

Seasonal and Celestial Energy Shifts

Beyond daily cycles, the changing of the seasons and planetary alignments affect how energy moves through the Earth.

Solstices: The Extremes of Light and Darkness

The solstices mark the longest and shortest days of the year. The Summer Solstice, with its abundant sunlight, is a time for celebration, growth, and expansion. The Winter Solstice, in contrast, is a period of introspection, stillness, and renewal.

Best practices for solstices:

Summer Solstice: Fire ceremonies, group meditations, and gratitude rituals for abundance.

Winter Solstice: Candlelit meditation, deep shadow work, and setting intentions for the new cycle.

Equinoxes: The Balance of Day and Night

Equinoxes symbolize equilibrium, where day and night are equal. These periods are powerful for harmonizing energy, rebalancing chakras, and embracing change.

Best practices for equinoxes:

Balance Poses in Yoga: Practicing Tree Pose or Eagle Pose enhances inner and outer balance.

Energy Clearing Rituals: Smudging spaces, fasting, or detoxing to refresh the system.

Nature Walks: Spending time in forests or near bodies of water enhances grounding.

Lunar Influence: The Moon's Magnetic Pull

The moon's phases significantly influence human emotions, tides, and energetic flow. Full moons amplify intentions, while new moons are ideal for new beginnings.

Best practices for lunar phases:

Full Moon: Charging crystals, moon bathing, and releasing old patterns.

New Moon: Setting goals, vision boarding, and planting seeds of manifestation.

By aligning spiritual practices with these celestial rhythms, one can deepen their connection to Earth's cycles and harness natural energy flows for greater transformation and insight.

MAPPING ENERGY:
FINDING YOUR OWN POWER SPOTS

A GUIDE FOR THE MODERN EXPLORER

Learn how to feel, measure, and interpret energy fields in nature

Bring your compass, EMF meter, or just trust your gut instincts. Discover practical tools and intuitive techniques to identify the world's hidden energy reservoirs.

Understanding the Landscape: Geological Clues to Energy Hotspots

The Earth itself provides the first clues to uncovering power spots. Geological formations rich in quartz, iron, or other conductive minerals often serve as amplifiers for natural electromagnetic energy. Areas with significant fault lines, underground water flow, or volcanic history tend to exhibit unique energy fields. When exploring, look for distinct rock formations, mineral veins, and geothermal activity as indicators of potential power spots.

Practical Tools for Energy Detection

1. **Compass:** A simple compass can be an invaluable tool for detecting geomagnetic anomalies. If the needle behaves erratically, swings in unusual directions, or struggles to settle, you may be standing in a location with an unusual magnetic field.

2. **EMF Meter:** Electromagnetic field meters are useful for measuring fluctuations in energy. High readings in natural areas could indicate an increased presence of telluric currents or underground minerals influencing the electromagnetic environment.

3. **Dowsing Rods:** Used for centuries to locate underground water sources, dowsing rods are also employed to detect energy fields. If the rods cross or move without external interference, it may indicate the presence of an energetic anomaly.

4. **Infrared Thermometers:** Some energy hotspots exhibit slight temperature variations due to localized magnetic or geothermal activity. Using an infrared thermometer can help track temperature shifts in the landscape.

5. **Sound and Resonance:** Certain power spots have unique acoustic properties, where echoes, resonance, or even an unusual silence contribute to their distinct energetic quality. Tuning forks and singing bowls can help test how sound interacts with the environment.

Developing Your Sixth Sense: Intuition and Sensory Awareness

While scientific tools provide valuable data, your own body is a powerful instrument for detecting energy. Many energy-sensitive individuals report tingling sensations, feelings of lightness, or emotional shifts when encountering high-energy sites. Tuning into these reactions can be just as effective as any device.

Techniques to Enhance Sensory Awareness:

Mindful Walking: Move slowly through the landscape, paying attention to bodily sensations and subtle environmental shifts.

Hand Sensitivity: Hold your hands just above rocks or soil and note any warmth, coolness, or vibrations.

Body Sway Test: Stand still with eyes closed; if you notice a gentle sway in one direction, it could indicate an energetic pull.

Signs from Nature: Observing Animals and Plant Growth

Nature itself responds to energy in fascinating ways. Plants often grow in unusual formations around power spots, and trees may twist or lean in unexpected directions. Animals, particularly birds and insects, tend to be drawn to these areas, while some locations may feel eerily devoid of life.

Reading Energy Maps and Aligning with Celestial Movements

Energy mapping involves tracking the alignment of power spots with celestial events such as solstices, equinoxes, and lunar cycles. Ancient cultures often built sacred sites in alignment with the stars, suggesting a connection between cosmic and terrestrial energies. Using topographic maps, satellite imagery, and historical data, modern explorers can identify potential hotspots aligned with celestial events.

Testing and Validating Power Spots

Once a potential power spot is identified, spending time in the area is key to determining its energetic significance. Try different spiritual practices, note any physical or emotional changes, and compare readings from scientific instruments. Keeping a journal of observations over time can help confirm whether an area consistently exhibits heightened energy.

How to Cultivate a Personal Connection to Power Spots

True understanding of power spots goes beyond detection—it involves forging a connection with the land. Spending time in these locations, offering gratitude, and approaching them with an open mind can deepen one's experience. Whether using meditation, chanting, or simple stillness, attuning to the space allows for greater absorption of its energy.

Ethical Exploration: Preserving and Protecting Energy Sites

Power spots are sacred, not only due to their energetic significance but also because they often hold cultural and historical value. Practicing ethical exploration by respecting the land, leaving no trace, and avoiding disruption ensures that these places remain vibrant for future seekers.

Final Thoughts: The Journey of Discovery

Mapping energy is a deeply personal journey that bridges science, intuition, and ancient wisdom. By combining modern tools with sensory awareness and respect for nature, anyone can uncover the hidden energy reservoirs that shape our world. Whether in the heart of the Rocky Mountains or an unassuming meadow, power spots are everywhere—waiting for those willing to listen.

HOW THE EARTH SPEAKS
THROUGH SUBTLE CUES AND SIGNALS

Recognizing synchronicities and natural indicators of high-energy zones

Feel the buzz underfoot, observe how animals behave, or notice sudden temperature changes—the Earth communicates in whispers if you know how to listen.

Tuning into the Earth's Subtle Signals

The Earth is alive with constant motion, subtle vibrations, and shifts in energy that affect not only landscapes but also those who are sensitive to its cues. Recognizing these signals requires patience, observation, and an open mind. Whether you are standing in the midst of a windswept valley, hiking through an ancient rock formation, or simply resting by a flowing stream, the environment is rich with messages waiting to be deciphered.

Feeling the Vibrations Beneath Your Feet

One of the most immediate ways to sense Earth's energy is by tuning into ground vibrations. Some power spots have unique geological features that create a distinct hum or resonance. These sensations may be the result of underground water movement, geothermal activity, or the piezoelectric properties of quartz-rich rock formations.

Barefoot Grounding: Walking barefoot on different surfaces can help attune your body to these vibrations. Sandstone formations, granite peaks, and volcanic rock all emit different frequencies that can be perceived with practice.

Resonance Testing: In locations known for heightened energy, some visitors report feeling a tingling sensation in their feet or hands, an indication of electromagnetic interaction with the body's own bioelectric field.

Observing Animal Behavior

Animals are highly attuned to environmental changes and can serve as natural indicators of energetic shifts. Birds, insects, and mammals often react to geomagnetic fluctuations long before humans notice them.

Bird Flight Patterns: Birds sometimes gather in unusual formations over energetically charged locations, or they may suddenly vacate an area before an atmospheric or seismic event.

Insect Movements: Bees, butterflies, and other pollinators are particularly drawn to magnetic anomalies. A sudden increase or absence of insect activity in a natural space may signal an energetic fluctuation.

Mammalian Instincts: Deer, foxes, and other mammals tend to avoid areas with unstable energy fields. If an area feels energetically "off," yet is devoid of wildlife, it may warrant further exploration.

Sensing Temperature Variations

Unexplained temperature shifts are common in areas of high energy. These can be due to geothermal activity, underground water sources, or interactions between different air masses.

Cold Spots: Sudden drops in temperature may indicate the presence of a vortex or an area where energy is concentrated.

Warm Zones: Some sacred locations have been noted to radiate warmth even in cooler weather, possibly due to their alignment with telluric currents or geothermal pockets.

Listening to the Sounds of the Land

Sound is another way the Earth communicates its energy. Some locations produce low-frequency hums, echoes, or even musical tones caused by wind passing through rock formations or underground structures.

Echo Chambers: Certain rock formations act as natural sound amplifiers, producing distinct reverberations. Testing these acoustics through chanting, drumming, or singing can be a powerful way to interact with the energy of the land.

Subterranean Rumbles: Some power spots sit above underground rivers, fault lines, or geothermal activity, creating deep, almost imperceptible sounds. Being still and listening carefully in these areas can reveal these Earth-born frequencies.

Cloud Formations and Atmospheric Shifts

The sky itself often mirrors the energy of the land below. Paying attention to cloud movement, formations, and weather shifts can help determine the activity of a power spot.

Lenticular Clouds: Often forming above high-energy locations, these lens-shaped clouds indicate strong wind currents interacting with the land's electromagnetic fields.

Sudden Wind Changes: A still day that suddenly becomes breezy or a location where wind patterns shift dramatically may signal an energetic anomaly.

Water's Role in Energy Mapping

Water, especially flowing or underground streams, plays a crucial role in amplifying and distributing Earth's energy.

Dowsing for Water: Using dowsing rods or simply observing how water behaves in an area can provide insights into subterranean energy flows.

Mineral Springs: Naturally occurring mineral springs are often found in locations rich in Earth energy. The high mineral content enhances conductivity and amplifies telluric currents.

Practicing Energy Mapping Techniques

By incorporating scientific instruments with natural observation, energy seekers can build a more complete picture of how the Earth speaks through subtle cues. Tools such as:

- Magnetometers to measure fluctuations in the magnetic field.
- Infrared thermometers to detect temperature anomalies.
- EMF meters to record electromagnetic activity.

Final Thoughts: Developing Your Sensory Awareness

The Earth's energy is constantly shifting, speaking in whispers that require patience and practice to interpret. By refining sensory awareness and trusting intuition, you can unlock the ability to read the signs of the land, allowing for deeper connection, greater respect, and a richer experience when visiting Colorado's power spots and beyond. The more you listen, the more the Earth reveals its secrets, drawing you into the rhythms and wisdom of the natural world.

HISTORICAL AND CULTURAL PERSPECTIVES

ECHOES OF THE FIRST GUARDIANS OF SACRED SPACES

Exploring Native American perspectives on Earth's energy and sacred places

For centuries, Indigenous tribes revered these locations as sacred. Their stories, rituals, and traditions offer profound lessons in respect and harmony with the land.

The Spiritual Connection
Between Indigenous Tribes and Sacred Spaces

Indigenous cultures across the world have long recognized the profound spiritual energy present in the land. In North America, the Indigenous tribes of Colorado, including the Ute, Cheyenne, Arapaho, and Apache, held deep spiritual connections to the landscapes they called home. They understood that the Earth itself was alive, resonating with energy, history, and wisdom.

Sacred places were not randomly chosen; they were identified through careful observation, intuitive understanding, and generational knowledge passed down through oral traditions. These locations were believed to be portals to other realms, places of healing, and areas where spirits resided. Indigenous tribes lived in harmony with these landscapes, holding ceremonies, making offerings, and ensuring that these sites remained unspoiled.

Land as a Living Entity: The Indigenous Worldview

For many Indigenous tribes, the land was—and still is—considered a sentient being, possessing its own consciousness and energy. The concept of land as a living entity shaped their spiritual practices, social structures, and ecological stewardship. This perspective is evident in the many sacred sites that were seen as places where the Earth spoke most clearly to those who listened. Mountains, rivers, canyons, and rock formations were considered the physical manifestations of spirits and ancestors, guiding tribes in their way of life.

Ceremonial Sites and Rituals

Sacred places were often the locations for important tribal ceremonies, including vision quests, healing rituals, and rites of passage. These ceremonies were designed to align the human spirit with the greater forces of nature, the cosmos, and the unseen world. Some of the most common rituals included:

Vision Quests: Individuals seeking guidance or spiritual insight would spend days fasting and meditating in isolated sacred locations, awaiting messages from the spirit world.

Sun Dances: Practiced by the Plains tribes, the Sun Dance was a ceremony of renewal, endurance, and sacrifice, performed in sacred spaces aligned with the movements of celestial bodies.

Healing Ceremonies: Shamans and medicine men would conduct rituals in energy-rich sites, using chants, drumming, and herbal medicine to harness the land's natural healing properties.

Offerings and Prayers: Gifts such as tobacco, beads, and food were offered to spirits in gratitude for guidance, protection, and sustenance.

Sacred Sites in Colorado

Several locations in Colorado have long been recognized by Indigenous tribes as sacred spaces. These places are not only geological wonders but also energetic focal points, where the veil between the physical and spiritual worlds is said to be thinner.

The San Luis Valley

The San Luis Valley, often described as one of the most mysterious places in North America, was considered a spiritual crossroads for many Indigenous tribes. It was a place of pilgrimage, where people from various nations would gather to perform ceremonies and seek wisdom from the spirits of the land. Today, the valley continues to be a site of high strangeness, with reports of unusual lights, UFO sightings, and unexplained phenomena.

The Great Sand Dunes

The Great Sand Dunes, rising abruptly from the valley floor, hold a powerful presence in Indigenous mythology. The Ute believed these dunes were created by the spirits and that they served as a place of great spiritual significance. Many tribes conducted ceremonies near the dunes, believing that the shifting sands carried messages from the ancestors.

Garden of the Gods

This striking red rock formation was considered a place of power and transformation by the Ute and Cheyenne. Legends speak of warriors and shamans receiving visions among the towering stones, where the Earth's magnetic fields are believed to create heightened spiritual awareness.

Indigenous Knowledge and Modern Spirituality

As interest in sacred landscapes grows, many non-Indigenous seekers are drawn to these sites for meditation, healing, and energy work. However, it is important to approach these places with reverence and respect, acknowledging the cultural and spiritual significance they hold for Indigenous communities. Learning from traditional wisdom can offer deeper insights into how to connect with these energies ethically and harmoniously.

Practicing mindfulness, leaving offerings of gratitude, and understanding the history of these spaces help preserve their sacredness for future generations.

Many Indigenous communities are working to protect these sites from encroachment, ensuring that their sacred places remain undisturbed.

Conclusion

The Indigenous wisdom surrounding sacred landscapes teaches us how to see the Earth as a living entity, deserving of care and respect. These sites are more than just places of historical significance; they are living, breathing power spots that continue to shape human consciousness. By recognizing and honoring the traditions of those who first walked these lands, we can deepen our own connection to the Earth's energy and wisdom, ensuring that these sacred spaces endure for generations to come.

ANCIENT MESSAGES ETCHED IN STONE

Understanding how ancient cultures recorded and honored power spots

Throughout Colorado and the greater Southwest, ancient Indigenous peoples left behind an enduring legacy in the form of petroglyphs and pictographs—carved and painted images that speak of their connection to the land, the cosmos, and the spiritual world. These artistic expressions, often found in rock shelters, canyon walls, and sacred sites, provide a fascinating glimpse into the lives, beliefs, and ceremonial practices of those who once inhabited these lands.

The Meaning and Significance of Rock Art

Petroglyphs, which are carved or pecked into rock surfaces, and pictographs, which are painted using natural pigments, have long been viewed as more than mere decorations. These images are thought to hold spiritual and symbolic significance, representing visions, celestial events, shamanic experiences, and messages to future generations. Many Indigenous traditions believe that these markings contain sacred energy and serve as portals to the spirit world.

Common symbols found in Colorado's petroglyphs and pictographs include:

Spirals: Often interpreted as symbols of energy flow, transformation, or celestial cycles.

Anthropomorphic Figures: Some figures appear to have elaborate headdresses or elongated limbs, potentially depicting shamans or spiritual beings.

Animals: Depictions of bighorn sheep, birds, snakes, and other wildlife often represent totemic connections or spiritual guides.

Handprints: Found in many sites, handprints may indicate a personal mark of presence, ceremonial initiation, or communication with the ancestors.

Sacred Locations with Petroglyphs and Pictographs

Several locations in Colorado are renowned for their ancient rock art, each carrying its own unique energy and historical significance.

Canyon Pintado (Painted Canyon)

Located in western Colorado, this site is home to thousands of rock art images left by the Fremont and Ute peoples. The intricate carvings and paintings depict hunting scenes, spiritual figures, and geometric patterns believed to hold cosmic or ceremonial meaning.

Picketwire Canyonlands

Best known for its extensive dinosaur tracks, Picketwire Canyon also contains rock art sites that date back thousands of years. Some images are believed to be shamanic in nature, possibly created during vision quests or trance states induced by the powerful energy of the land.

Picture Canyon

This remote canyon in southeastern Colorado contains well-preserved pictographs created by Plains tribes. The vibrant ochre and charcoal paintings are thought to depict stories of migration, ceremonial dances, and connections with celestial beings.

Shavano Valley Rock Art Site

One of the most significant Ute rock art sites in Colorado, Shavano Valley holds petroglyphs that tell a rich history of storytelling, hunting, and shamanic practices. The energy of this place is palpable, as many visitors report experiencing a deep sense of reverence while standing before the ancient carvings.

The Energetic and Spiritual Power of Rock Art

Many power spots in Colorado are home to petroglyphs and pictographs, reinforcing the idea that these locations were chosen for their energetic significance. Indigenous shamans and spiritual leaders likely perceived the heightened electromagnetic activity, telluric currents, or celestial alignments in these places, making them ideal for sacred ceremonies and spiritual communication.

Visitors to these sites often describe feeling a shift in awareness, as though stepping into an ancient frequency preserved through time. The resonance of these locations encourages deep meditation, reflection, and an enhanced connection to the past.

Preserving the Sacred Legacy of Rock Art

As more people become interested in exploring these ancient sites, it is crucial to protect and respect them. Many petroglyphs and pictographs have suffered damage from vandalism, weathering, and human interference. Conservation efforts by Indigenous communities, archaeologists, and preservation groups are working to ensure that these sacred messages remain intact for future generations.

To visit rock art sites respectfully:

Never touch the carvings or paintings. Oils from human skin can accelerate the degradation process.

Stay on designated trails. Foot traffic can disturb the fragile ecosystems around these sacred spaces.

Do not leave modern markings. Carving or painting over ancient rock art is not only disrespectful but also illegal.

Learn about the site's history before visiting. Understanding the cultural significance of a place enhances the experience and fosters greater appreciation.

By honoring these ancient expressions, we contribute to the preservation of the wisdom and spiritual traditions of the First Peoples, ensuring that their voices continue to be heard through the echoes of stone.

Petroglyphs and pictographs are more than relics of the past; they are living testaments to the deep relationship Indigenous peoples have with the land and the unseen forces that shape our world. As we explore these sacred spaces, we are invited to listen, learn, and reflect on the timeless messages etched into the very fabric of the Earth.

COLORADO'S TRANSFORMATION

The impact of human activity on natural energy fields and sacred landscapes

Colorado's mining boom didn't just dig for gold; it unearthed energy corridors that still resonate today. Explore the scars, the stories, and the lingering energy they left behind.

The Birth of Colorado's Mining Industry

The discovery of gold in the mid-19th century ignited a frenzy that transformed Colorado's landscape. From the rugged peaks of the Rockies to the deep canyons of the San Juan Mountains, thousands of prospectors flooded the region, carving tunnels and shafts deep into the Earth. While many came seeking fortune, what they left behind was a network of mineral-rich veins, vast excavation sites, and energetic imprints that remain active today.

The Gold and Silver Rush: A Transformation of Land and Energy

The Pike's Peak Gold Rush (1858-1861) marked the beginning of Colorado's mining era. The rush led to the establishment of booming mining towns such as Central City, Leadville, and Cripple Creek. As prospectors extracted gold and silver, they inadvertently disrupted the natural energy fields of the land, exposing vast deposits of conductive minerals like quartz, iron, and copper.

Gold and Quartz: The Energetic Connection

Gold is often found in quartz veins, a mineral known for its piezoelectric properties. When pressure is applied, quartz generates an electric charge, which could explain the lingering energy that many visitors experience at old mining sites. Some spiritual seekers believe that the residual energy from mining operations enhances the natural electromagnetic properties of these locations, making them more conducive to meditation and energetic work.

Mining Towns and Ghostly Echoes

Mining towns flourished and then faded, leaving behind eerie remnants of once-thriving communities. Places like St. Elmo, Silverton, and Victor became ghost towns, where abandoned buildings and old mines stand as silent witnesses to the past. Paranormal investigators and energy workers often report heightened spiritual activity in these areas, suggesting that the emotional intensity and hardship of the mining era left behind an energetic imprint.

The Industrialization of Mining: Tunnels, Railroads, and Power Disruptions

With the arrival of industrial-scale mining in the late 19th and early 20th centuries, the landscape was further transformed. The construction of railroads, processing mills, and smelting operations introduced large-scale mechanical disruptions to the Earth's energy fields.

Railroads and the Shifting Energy Grid

The Denver & Rio Grande Western Railroad and other major rail lines played a key role in transporting minerals from remote locations to processing centers. However, the extensive railway network also affected the energy landscape by altering natural ley lines—geometric alignments of energy that some believe are key to Earth's vibrational structure.

The Impact of Heavy Metal Extraction

Extracting metals such as lead, zinc, and copper disrupted natural conductivity patterns in the Earth. Some spiritual practitioners believe that these disturbances contributed to energetic blockages in certain areas, while others suggest that they uncovered hidden energy channels that had previously remained dormant.

Environmental and Energetic Scars of the Mining Era

Mining operations left behind more than just abandoned tunnels and slag heaps; they created long-lasting environmental challenges. Acid mine drainage, heavy metal contamination, and soil erosion continue to affect the land. Yet, some believe that these areas, while damaged, have become sites of energetic transformation.

The Reclamation of Energy

Some abandoned mines have naturally transitioned into sites of regrowth. Trees, moss, and flowing water now reclaim landscapes once dominated by industry. Water flowing through old mines often carries minerals that contribute to increased conductivity, potentially revitalizing the energetic flow of the land.

The Role of the EPA and Superfund Sites in Restoring Energy Balance

Mining, while foundational to Colorado's economy and history, left behind extensive environmental damage. Recognizing the severe impact of industrial mining, the Environmental Protection Agency (EPA) established the Superfund program to clean up contaminated sites across the country, including several in Colorado. These efforts are essential not only for environmental restoration but also for rebalancing the energy of these disturbed landscapes.

The EPA's involvement in addressing Colorado's most toxic sites, such as the Summitville Mine, the California Gulch in Leadville, and the Rocky Mountain Arsenal, demonstrates the long-term consequences of unchecked mining and industrial pollution. These areas suffered extensive contamination from heavy metals, acidic runoff, and hazardous chemicals. Despite the restoration work, many believe that the scars on the land also left energetic imprints that require spiritual healing.

Modern energy workers and healers have taken an interest in these sites, recognizing that beyond physical restoration, these locations may need energetic recalibration. The belief is that the land holds onto trauma—both ecological and human-induced—and that targeted energy work can assist in clearing these disruptions. Some of the techniques employed include:

Crystal Grid Work: Healers place specific stones such as black tourmaline, quartz, and hematite at key points within a site to transmute stagnant energy and restore the natural energetic flow of the land.

Sound Healing and Vibrational Therapy: Using Tibetan singing bowls, tuning forks, and chanting, practitioners aim to break up negative energetic imprints left by industrial exploitation.

Reiki and Earth Healing Ceremonies: Energy healers perform group ceremonies that involve meditative intention-setting, calling upon the Earth's regenerative forces to aid in renewal.

Water Blessings and Offerings: Since many contaminated sites involve polluted waterways, spiritual activists conduct rituals to purify the water energetically, sometimes in tandem with scientific water-cleaning efforts.

One of the biggest concerns in the environmental and spiritual communities is the long-term energetic health of these areas. Even though the EPA's physical cleanup processes mitigate environmental hazards, the belief among spiritual practitioners is that deep-seated energetic wounds remain. The goal is to complement government-led restoration efforts with holistic healing practices, bringing balance back to the land.

As more awareness is brought to the impact of industrial activity on natural energy systems, collaborations between scientists, Indigenous groups, and spiritual healers are forming. These partnerships focus on restoring both the physical and spiritual integrity of former mining sites, ensuring that these spaces can transition from scars of exploitation to areas of renewal and revitalization.

Colorado's transformation from an industrial hub of mining to a landscape undergoing deep healing presents a powerful opportunity for energy work. By respecting both scientific restoration methods and the ancient wisdom of those who understood the land's spiritual nature, we can help reclaim these sites—not just as places of history, but as power spots for the future.

Modern Energy Workers and Healers

In recent years, spiritual practitioners have sought to heal these sites by conducting energy clearing rituals, crystal grid activations, and meditation retreats in former mining towns. The belief is that these areas, once disrupted by industry, can be realigned with Earth's natural energy network through conscious intention and restorative practices.

The Future of Colorado's Mining Legacy

Today, many former mining sites have been transformed into historical landmarks, hiking destinations, and even centers for spiritual exploration. Locations such as the Crystal Mill in Marble, the Argo Gold Mill in Idaho Springs, and the Mollie Kathleen Mine in Cripple Creek attract visitors seeking both history and energy experiences.

While the scars of mining remain, the land continues to evolve, finding new ways to restore balance. By acknowledging both the historical and energetic impacts of mining, we can appreciate Colorado's transformation—not just as a story of industry, but as an ongoing journey of reclamation and renewal.

BEYOND THE PHYSICAL: EXPLORING INTUITION AND CONNECTION

WRITING THE ENERGY YOU FEEL
AND BUILDING INNER CLARITY

How documenting your experiences can accelerate your personal growth

Track your experiences and impressions at power spots. Journaling can deepen your understanding of the sensations and thoughts that arise in these profound spaces.

The Power of Journaling in Energy Exploration

Journaling is a bridge between experience and understanding, a way to make sense of the unseen forces that shape our perceptions. When visiting power spots, the act of writing helps clarify emotions, record intuitive impressions, and document personal transformations. It serves as both a scientific and spiritual tool, capturing the energies felt, the physical sensations encountered, and the thoughts that emerge in heightened states of awareness.

Setting Intentions Before Journaling

Before visiting a power spot, take a moment to set an intention for your journaling practice. What do you hope to discover? Are you seeking guidance, emotional clarity, or simply an open-ended experience? Setting an intention frames your experience and primes your subconscious mind to be receptive to messages from the land.

Some sample intentions include:

Seeking clarity on a life decision – Allow the power spot's energy to influence your thoughts and emotions.

Understanding personal energy shifts – Notice how different locations affect your physical and mental state.

Developing intuition – Use journaling to record moments of heightened awareness, synchronicities, or sudden insights.

Techniques for Energy Journaling

1. **Freewriting:** This technique involves writing without stopping or censoring your thoughts. Let words flow without concern for structure or grammar. Describe the sensations you feel in your body, the emotions that arise, and any intuitive messages you receive from the land.
2. **Sensory Mapping:** Create a map of your surroundings by noting sensory experiences in different areas of a power spot. Document changes in temperature, sounds, scents, and the feeling of the earth beneath you. Does one area feel more charged or peaceful than another?
3. **Emotional Reflections:** Pay attention to shifts in your mood and energy levels. Do you feel lighter, more grounded, or overwhelmed? Some power spots evoke feelings of deep serenity, while others stir up old emotions or offer a sense of renewal. Use your journal to explore these emotional layers.
4. **Sketching and Symbolism:** Words aren't the only way to capture an experience. Sketching landscapes, drawing symbols, or writing poetry inspired by the energy of a place can help encode your experience in a creative, non-linear way.
5. **Comparing Locations Over Time:** Revisit your journal after multiple visits to the same power spot. Are there recurring themes? Has your perception of the place changed? This longitudinal approach can reveal patterns in your energy interactions.

The Science Behind Writing and Reflection

Writing has been shown to activate neural pathways associated with memory, emotion, and problem-solving. When journaling about energy experiences, this cognitive engagement enhances recall and deepens personal insights. The act of reflection solidifies intuitive downloads, making them more tangible and actionable.

Honoring the Experience Through Writing

Approach your journal with respect, treating it as a sacred record of your spiritual and energetic journey. Some practitioners create dedicated notebooks for power spot visits, incorporating pressed leaves, dried flowers, or small sketches to honor the essence of a place.

Closing Thoughts: Integrating Insights

Journaling doesn't end when you leave a power spot—it continues as an evolving dialogue between you and the energy of the Earth. Reflecting on past entries allows you to integrate lessons, recognize growth, and cultivate deeper awareness of how these experiences shape your understanding of energy and connection.

UNLOCKING AWARENESS AND EMBRACING EARTH'S ENERGY

Developing sensitivity to the natural forces that surround us

Heighten your perception to experience the world's subtle energies on a deeper level. With practice, anyone can unlock this transformative skill.

Tuning into the Earth's energy requires a blend of mindfulness, patience, and practice. Our senses are constantly engaged, but learning to attune them to subtler vibrations can enhance awareness and deepen the connection to natural power spots. By refining sensory perception, individuals can detect shifts in energy, recognize patterns in nature, and experience an amplified sense of presence in sacred spaces.

One effective way to enhance sensitivity is through focused observation. Begin by immersing yourself in a natural environment and taking note of small details—the movement of leaves, temperature variations, the way sound travels in different spaces. The more you pay attention, the more attuned you become to subtle shifts in your surroundings.

Practicing energy scanning with your hands can also enhance awareness. Holding your hands a few inches above different surfaces—rocks, water, or trees—may reveal warmth, tingling, or pulsations, indicating varying energy fields. Combining this with deep, rhythmic breathing can help synchronize your body's energy with the environment.

By tuning into the Earth's subtle signals, we gain access to a profound sense of interconnectedness, allowing us to navigate sacred landscapes with deeper awareness and appreciation.

PRACTICAL TIPS
FOR VISITING POWER SPOTS

SURVIVAL MEETS SPIRIT
WITH ESSENTIAL TOOLS FOR EVERY JOURNEY

What to bring and how to prepare for deep energy experiences

Pack smart and intentionally. From boots to energy-amplifying crystals, ensure you're prepared for both physical and spiritual exploration.

When venturing into power spots, a well-prepared traveler can enhance both their physical safety and their ability to connect with the land's energy. Whether hiking to a remote mountain vortex or meditating at a sacred valley, having the right tools ensures a smooth and enriching experience.

Essential Gear for Physical Comfort and Safety

Sturdy Footwear: Uneven terrain and rocky paths require reliable hiking boots to prevent injuries.

Layered Clothing: Weather conditions can change rapidly, so dressing in layers allows for adaptability.

Water and Snacks: Hydration and sustenance are crucial, especially in high-altitude locations.

Navigation Tools: A physical map, compass, or GPS ensures safe exploration in remote areas.

First Aid Kit: Minor injuries can happen, so carrying essential medical supplies is always wise.

Spiritual and Energetic Tools

Crystals: Stones like clear quartz, amethyst, and black tourmaline amplify and balance energy.

Journal and Pen: Recording insights, feelings, and messages from nature deepens personal reflection.

Sage or Palo Santo: Used for energy cleansing, these tools help clear stagnant energy before meditation.

Singing Bowl or Chimes: Sound vibrations enhance the energetic frequency of a place.

By preparing mindfully, visitors can fully immerse themselves in the energy of power spots, experiencing both physical adventure and deep spiritual connection.

Ethical Practices:
Honor the Energy, Leave No Trace, and Protect Sacred Spaces

How to explore these locations with respect and mindfulness?

Respect the land. Preserve its wild beauty and electrifying energy for future generations.

Power spots are sacred, not just for their natural beauty but for their energetic and historical significance. Practicing ethical exploration ensures these places remain undisturbed and vibrant for years to come. Every visitor carries a responsibility to honor the land, leaving it as they found it—or better.

Leave No Trace Principles

Following the principles of Leave No Trace is essential for visiting sacred sites:

Pack It In, Pack It Out: Always carry away all trash, food waste, and personal belongings.

Stay on Marked Trails: Prevent erosion and habitat destruction by avoiding off-trail excursions.

Respect Wildlife: Observe animals from a distance and never feed or disrupt them.

Honoring the Energy of Sacred Spaces

Beyond physical conservation, energy preservation is equally important. Be mindful of your presence and actions:

Avoid Disrupting Ritual Sites: Many sacred locations hold deep cultural significance. Do not alter or remove offerings, stones, or artifacts.

Practice Quiet Reflection: Loud noises, shouting, or playing music can disturb the natural energy balance.

Use Eco-Friendly Rituals: If performing ceremonies, use natural, biodegradable materials.

By practicing mindfulness and respect, we ensure that future generations can continue to experience the power and beauty of these sacred sites.

THE FUTURE
OF COLORADO'S POWER SPOTS

HOW MODERNITY CHALLENGES SACRED SPACES

The environmental and societal risks facing these power spots

Urban sprawl and climate change threaten these sacred sites. What's at stake if we lose them, and how can we prevent it?

As Colorado's cities expand and the effects of climate change intensify, the state's power spots face mounting threats. The very landscapes that have been revered for their energetic, cultural, and geological significance are increasingly encroached upon by human development. What was once untouched land is now at risk of being paved over, fragmented, or disrupted by industrialization. Without intervention, the balance of these sacred spaces may be permanently altered.

Urbanization and the Displacement of Power Spots

As urban areas like Denver, Boulder, and Colorado Springs grow, the natural spaces surrounding them shrink. Highways, housing developments, and commercial expansion encroach on regions historically known for their energy fields. The more we build, the more we displace these power spots, breaking the energetic grid that once flowed uninterrupted through mountains, valleys, and rivers.

The construction of roads and infrastructure disrupts the subtle electromagnetic fields present in these sites. Power spots located near geothermal activity, rock formations with high quartz content, and mineral-rich deposits may see a decline in their natural energy signatures due to soil disturbance and vibration from heavy machinery. These places may lose their resonance as the earth's energy lines are severed by human activity.

Climate Change and the Alteration of Natural Energy Flows

Climate change is not just about rising temperatures—it's also about shifting weather patterns, increased wildfires, and disrupted ecosystems. Many of Colorado's most sacred locations depend on stable environmental conditions to maintain their energy balance. Places like Mount Princeton Hot Springs, where geothermal activity contributes to the region's energy field, are vulnerable to changes in groundwater flow caused by prolonged droughts or excessive human water use.

Glacial and permafrost melting also contribute to shifts in energy patterns. The Rocky Mountains' unique ability to store and distribute energy may be compromised as ice sheets retreat, altering the natural flow of water and the energetic properties of the land.

Tourism, Overuse, and the Energetic Drain

As power spots become more well-known, the influx of visitors can unintentionally degrade their energy. Foot traffic, littering, and human interference take a toll on these places, slowly wearing down their vibrational intensity. Overcrowding at locations such as Garden of the Gods and the San Luis Valley leads to soil erosion, vegetation loss, and a diminished sense of solitude—key components in maintaining a site's energetic integrity.

Many spiritual seekers visit these locations to meditate, recharge, and connect with the Earth's energy. However, without proper education on conservation, well-intentioned visitors may unintentionally drain the space rather than rejuvenate it. Sites need periods of rest, just as the human body does, in order to sustain their energetic properties.

How Can We Protect These Sacred Spaces?

1. Conservation and Land Preservation: Organizations and local communities must work to acquire and protect key energy sites from development. Land trusts and conservation easements can help ensure these places remain untouched for future generations.

2. Sustainable Tourism: Implementing visitor caps, designated pathways, and conservation education can mitigate damage while still allowing people to experience these powerful locations.

3. Eco-Friendly Infrastructure: Where development is necessary, using environmentally sensitive construction techniques can minimize disruption to natural energy grids.

4. Energy Restoration Practices: Modern energy workers and Indigenous communities can collaborate to conduct land blessings, cleansing ceremonies, and rebalancing efforts to heal areas impacted by human activity.

By understanding what is at stake and taking conscious action, we can help preserve Colorado's power spots before they become relics of the past. These sites are not just geological wonders—they are living, breathing centers of energy that deserve protection, reverence, and mindful stewardship.

The Earth Calls—Will You Answer?
Transforming Knowledge into Action

Colorado's places of power aren't just geological wonders; they're invitations to awaken, reconnect, and transform. Are you ready to heed the call and embrace the mystical edge of the natural world?

This book was only the beginning.

The landscapes you've explored, the currents you've touched,
and the reflections you've found are alive—changing with the seasons,
shifting with your steps.

Continue your journey with the Places of Power Interactive Map.

Follow new trails, revisit old pathways with new eyes,
and stay connected to the living energy of the land.

The Earth is still speaking—and you are part of the conversation.

📱 Scan the QR Code or visit
https://tinyurl.com/mwxzt3jj
to keep exploring.

GLOSSARY OF ENERGY & GEOLOGY TERMS

Energy & Vibrational Terms

Alignment (Energetic)
A state in which one's body, breath, and mind resonate with the energetic frequency of a place or practice.

Alpha Waves
Brainwave frequency (8–13 Hz) associated with calm, relaxed states—often experienced during meditation or moments of stillness in nature.

Atmospheric Ionization
The presence of charged particles (ions) in the air, often increased near waterfalls, after storms, or at high elevation. Known to influence mood and physiological response.

Bioelectric Field
The electromagnetic field naturally emitted by the human body, especially the heart and brain. Can be influenced by external geophysical conditions.

Chakra
Energy centers in the body (from yogic traditions) associated with specific physical, emotional, and spiritual functions. Often aligned with specific frequencies or natural elements.

Conduction (Energetic)
The movement of subtle energy through materials or landscapes—often heightened in places with mineralized rock, flowing water, or electromagnetic variation.

Energy Signature
The unique vibrational quality of a place, determined by its geology, water flow, atmospheric conditions, and symbolic presence.

Grounding
A practice of physically or energetically connecting with the Earth, often done barefoot or with attention to sensation, to calm the nervous system and align the body's energy field.

Ley Line
Hypothetical alignments between sacred sites, geological features, or energy hotspots believed to channel subtle Earth energy.

Piezoelectric Effect
The generation of an electrical charge when certain crystals (like quartz) are placed under mechanical stress or pressure. May influence the energetic feel of a location.

Telluric Currents
Subtle natural electric currents that flow through the Earth's crust. These may concentrate around conductive rock, water movement, or fault zones.

Third Eye Activation
A sensation or state of heightened intuition and perception, often associated with the pineal gland and crown of the head—linked to vibrational clarity or energetic elevation.

Vibrational Sensitivity
The ability to perceive subtle energetic changes in a place or environment, often experienced as tingling, mood shifts, or intuitive knowing.

Geological & Physical Terms

Ash-Flow Tuff
A type of rock formed from volcanic ash compacted and welded together. Found in places shaped by ancient explosive eruptions, often highly porous and sculpted by erosion.

Caldera
A large volcanic crater formed by the collapse of land following a major eruption. Often associated with high geothermal and mineral activity.

Cirque
A bowl-shaped depression formed by glacial erosion, typically hosting alpine lakes and surrounded by steep cliffs. Natural resonance chambers.

Conductivity (Geological)
The ability of rocks, minerals, or soils to transmit electric current. Enhanced by moisture, metallic minerals, and clay-rich formations.

Continental Divide
A natural boundary separating watersheds that drain into different oceans. Energetically seen as a symbolic spine of high-altitude clarity and transition.

Feldspar
A common mineral found in granite and other igneous rocks. May enhance reflectivity and mineral charge in high-altitude zones.

Fault Zone
A fracture or break in the Earth's crust along which movement has occurred. Often associated with conductivity discontinuities and piezoelectric activity.

Glacial Till
Unsorted material left behind by melting glaciers, including rock, sand, and clay. Can affect local conductivity and soil chemistry.

Granite
A coarse-grained igneous rock rich in quartz and feldspar. Forms the foundation of many Colorado mountain ranges and is known for its grounding and stabilizing energy.

Gneiss & Schist
Metamorphic rocks formed under intense pressure and heat. Characterized by banding and mineral alignment, contributing to grounding and energetic density.

Ion Exchange
A chemical process in which minerals in rock or soil trade ions with water. Affects the energetic "feel" of groundwater and enhances conductivity.

Laramide Orogeny
A period of intense mountain-building that created the Rocky Mountains. Shaped many of Colorado's energy-rich geologic structures.

Quartz
A piezoelectric crystal found in many Colorado formations. Known to amplify energy, carry charge, and hold vibration.

Rift Zone
An area where the Earth's crust is being pulled apart, creating faults and often geothermal activity. Associated with high telluric current movement.

Sedimentary Layering
Geologic layers formed by the deposition of mineral and organic particles over time. Often visible in canyon walls and ridges—each layer holds geologic memory.

Travertine
A porous rock formed by the precipitation of calcium carbonate from mineral springs. Creates natural terraces and energetic "soft zones" in water-fed sites.

Uplift
The vertical rise of Earth's crust due to tectonic forces. Creates elevation, exposes ancient rock, and opens energetic fields at high altitudes.

These works inspired or informed many of the ideas in this book. Some offer grounding in geology and geophysics, others explore subtle energy, ancient technology, or consciousness studies. Whether you're looking to dive deeper into Earth's physical processes or explore its vibrational mysteries, the following resources provide rich terrain to explore:

Foundational Energy & Earth Resonance

Burke, John, and Kaj Halberg. Seed of Knowledge, Stone of Plenty: Understanding the Lost Technology of the Ancient Megalith-Builders. Council Oak Books, 2005.
A groundbreaking look at how ancient sacred sites may have harnessed natural electromagnetic energy—based on direct field measurements and cross-cultural patterns.

Becker, Robert O., and Gary Selden. The Body Electric: Electromagnetism and the Foundation of Life. Harper, 1985.
A classic text exploring the body's bioelectric systems and how external fields influence healing and consciousness.

Tiller, William A. Science and Human Transformation: Subtle Energies, Intentionality and Consciousness. Pavior Publishing, 1997.
Explores the interface between measurable science and human intention, especially in relation to energy fields and consciousness.

Geology, Geophysics, and Natural Energy

Montgomery, Carla. Environmental Geology. McGraw-Hill Education, 10th ed., 2013.

A broad but accessible textbook for understanding Earth's physical processes, including fault zones, rift systems, and groundwater movement.

Skinner, Brian J., and Stephen C. Porter. The Dynamic Earth: An Introduction to Physical Geology. Wiley, 5th ed., 2004.
Covers mineralogy, plate tectonics, and the geophysical forces shaping landscapes—useful for understanding conductivity and uplift.

Murck, Barbara W., et al. Visualizing Geology. Wiley, 3rd ed., 2010.
A graphically rich look at Earth systems, excellent for visual learners exploring sedimentary layers, folds, and volcanic origins.

Consciousness, Earth Mysteries & Sacred Geography

Devereux, Paul. Earth Memory: Sacred Sites – Doorways into Earth's Mysteries. Vega, 2002.
Explores the energetic qualities of sacred sites, geomancy, and how humans have interacted with Earth's energy over millennia.

Watkins, Alfred. The Old Straight Track. Abacus, 1925.
The foundational text on **ley lines**, proposing that ancient alignments between sacred sites represent a lost understanding of Earth energy.

Brennan, Barbara Ann. Hands of Light: A Guide to Healing Through the Human Energy Field. Bantam, 1987.
While more focused on the human aura, this book explores how energy fields interact with space and environment.

Pennick, Nigel. The Ancient Science of Geomancy: Living in Harmony with the Earth. Destiny Books, 1995.
A practical and philosophical look at land energy, alignments, and earth-based design principles.

c.d. is a Colorado native, lifelong learner, and passionate explorer of the unseen forces woven through stone, sky, and silence. With degrees in **English Writing and Literature**, and a current pursuit of a **degree in Applied Geology**, she brings a rare synthesis of poetic insight and scientific curiosity to her work.

Formerly a teacher and forever a student of place, she has spent years walking the ridgelines, hot springs, fault zones, and flower-choked valleys of the Rockies—**mapping not just the land, but the energy it holds**. Her writing bridges disciplines, inviting readers to feel geology with their whole bodies and to experience landscapes as living systems of memory, frequency, and transformation.

When she's not writing or metaphorically rock-hammering her way through outcrops, c.d. can be found **chasing thunderstorms, teaching with heart, practicing barefoot breathwork in alpine meadows, or plotting the next sacred detour**. Her work is driven by wonder, grounded in science, and open to whatever mystery walks beside her on the trail.

Have you visited one of these places and felt something shift?
Have a story, insight, or energy experience to share?
Let's keep the conversation going.

📷 **Follow along on Instagram**: @powerspotscolorado
I post trail insights, location updates, behind-the-scenes writing,
and new discoveries from Colorado's energetic landscapes.

📫 **Email me**: powerspotscolorado@gmail.com
I'd love to hear how these places have moved you—or to answer questions
about geology, energy, or crafting your own sacred routes.

🌀 Whether you're a fellow seeker, earth-based healer, scientist,
skeptic, or simply curious explorer—you're welcome here.

We walk the same land.
We breathe the same charge.
Let's stay connected.

www.ingramcontent.com/pod-product-compliance
Lightning Source LLC
Chambersburg PA
CBHW061751120626
46550CB00005B/1961